BASIC MOTOR CONTROL

BASIC MOTOR CONTROL

BY

AMELIA NGUYEN
CHLOE FORD
JIMI WHITE

KNOWLEDGE

KNOWLEDGE BAKERS

ISBN: 9789390013364 (HB)

Published by : **Knowledge Bakers**
Pune, India
Email: knowledgebakers2019@gmail.com

Digitally Printed at : **Replika Press Pvt. Ltd.**

PREFACE

Power distribution systems used in large commercial and industrial applications can be complex. Power may be distributed through switchgear, switchboards, transformers, and panelboards. Power distributed throughout a commercial or industrial application is used for a variety of applications such as heating, cooling, lighting, and motor-driven machinery. The contactor's contacts are closed to start the motor and opened to stop the motor. This is accomplished electromechanically using start and stop push buttons or other pilot devices wired to control the contactor. The overload relay protects the motor by disconnecting power to the motor when an overload condition exists. Although the overload relay provides protection from overloads, it does not provide short-circuit protection for the wiring supplying power to the motor. For this reason, a circuit breaker or fuses are also used. In many commercial and industrial applications, quite a few electric motors are required, and it is often desirable to control some or all of the motors from a central location. The apparatus designed for this function is the motor control center (MCC). A combination starter is a single enclosure containing the motor starter, fuses or circuit breaker, and a device for disconnecting power. Other devices associated with the motor, such as pushbuttons and indicator lights may also be included.

Electric motors are widely used in several electronic applications; to the point that we can say they are everywhere. In our homes we have multiple examples, such as fans, hair dryers, fan heaters, kitchen appliances and more. If we then consider a car, we can soon discover that there are different types of electric motors inside a vehicle: heating system, radiator cooling fans, electric windows, electric mirrors, seat control and much more.

The present book covers a comprehensive range of topics related to electrical motor control systems. The first part of the book defines terms and concepts related to motor control systems. The second part of the book focuses on pilot devices, such as switches and pushbuttons, common pilot devices, and zero-speed switches. It includes self-test questions to help readers assess their understanding of the material. The third part of the book discusses motor starters and contactors, including overload relays, manual motor starters, magnetic-motor starters, and control relays. The fourth part of the book discusses diagrams, such as schematic vs. wiring diagrams, numbering systems, conduit fill, and motor-lead connections. The fifth part of the book covers circuits, such as power circuits, control circuit's source of supply, automatic control circuits, and sump-pump circuits and the last part of the book discusses troubleshooting motor control systems. This book is an excellent reference for students, engineers, and technicians who work with motor control systems and valuable resource for anyone who wants to learn about electrical motor control systems.

We are grateful to all those persons as well as various books, manuals, periodicals, magazines, journals etc. that helped in the preparation of this book. In spite of the best efforts, it is possible that some errors may have occurred into the compilation and editing of the book. Further queries, constructive suggestions and criticisms for the improvement of the book are always welcomed and shall be thankfully acknowledged.

Amelia Nguyen
Chloe Ford
Jimi White

CONTENTS

I

Terms and Definitions

Learning Objectives

After studying this section, students should be able to:

- Describe common terms used in electrical circuits, such as voltage and power.
- Describe Ohm's law and Watt's law.
- Differentiate between low-voltage release (LVR) and low-voltage protection (LVP).
- Determine common ratings used on overcurrent and overload devices.
- Distinguish between single-phase circuits and three-phase circuits.
- Be familiar with the terms plugging and anti-plugging.

1.

Ohm's Law and Watt's Law

This section provides a brief description of two of the most fundamental electrical relationships: **Ohm's law**, which describes current flow, and **Watt's law**, which describes how power is dissipated.

Click play on the following audio player to listen along as you read this section.

A BCcampus element has been excluded from this version of the text. You can view it online here: https://pressbooks.bccampus.ca/basicmotorcontrol/?p=21 (https://pressbooks.bccampus.ca/basicmotorcontrol/?p=21#pb-interactive-content)

Ohm's Law

Combining the elements of **voltage**, **current**, and **resistance**, George Ohm developed the following formula:

$$I = \frac{E}{R}$$

Where

- E = Voltage in volts
- I = Current in amps
- R = Resistance in ohms

This is called Ohm's law.

Let's say, for example, that we have a circuit with the potential of 1 volt, a current of 1 amp, and resistance of 1 ohm. Using Ohm's law we can say:

$$1A = \frac{1V}{1\text{ ohm}}$$

Let's say this represents a tank with a wide hose. The amount of water in the tank is defined as 1 volt, and the “narrowness” (resistance to flow) of the hose is defined as 1 ohm. Using Ohm’s law, this gives us a flow (current) of 1 amp.

Using this analogy, let’s now look at the tank with the narrow hose. Because the hose is narrower, its resistance to flow is higher. Let’s define this resistance as 2 ohms. The amount of water in the tank is the same as the other tank, so, using Ohm’s law, our equation for the tank with the narrow hose is:

$$? = \frac{1V}{2 \text{ ohms}}$$

But what is the current? Because the resistance is greater and the voltage is the same, this gives us a current value of 0.5 amps:

$$0.5A = \frac{1V}{2 \text{ ohms}}$$

Watt’s Law

Electric **power** is the rate at which energy is transferred. It’s measured in terms of joules per second (J/s). One joule of work done every second means that power is dissipated at a rate equal to one **watt (W)**.

Given the few basic electricity terms we know, how could we calculate power in a circuit?

Well, we have a standard measurement involving electromotive force, also know as **volts (E)**.

Current, another of our favourite electricity terms, measures charge flow over time in terms of the **ampere (A)**, which equals 1 coulomb per second (C/s). Put the two together, and what do we get? Power!

To calculate the power of any particular component in a circuit, multiply the voltage drop across it by the current running through it.

For instance, if current flows at a rate of 10 amps while voltage is 10 volts, then the circuit dissipates power at a rate of 100W.

2.

Low-Voltage Release vs. Low-Voltage Protection

Click play on the following audio player to listen along as you read this section.

A BCcampus element has been excluded from this version of the text. You can view it online here: https://pressbooks.bccampus.ca/basicmotorcontrol/?p=24 (https://pressbooks.bccampus.ca/basicmotorcontrol/?p=24#pb-interactive-content)

Two of the first terms that we will cover are low-voltage release (LVR) and low-voltage protection (LVP).

Sometimes called under-voltage release, **low-voltage release (LVR)** is a property that circuits have when upon a return of **voltage** following a power outage, loads automatically turn back on.

Sometimes called under-voltage protection, **low-voltage protection (LVP)** is a property that circuits have when upon a return of voltage following a power outage, loads will not automatically turn back on and will require further input from the operator.

A simple example of low-voltage release (LVR) and low-voltage protection (LVP) is a simple lighting circuit and a household microwave. Imagine that you are at home, heating something in the microwave (perhaps a burrito) when all of a sudden all of the lights go out. But not just the lights, the microwave and every electrical device not powered by a battery is dead.

Perhaps a tree has fallen across some power lines? Regardless of the cause, the effect is the same: the power is out. This is a common and annoying experience for us all, and often there is nothing to do but light some candles, read a book, and wait.

How do we know when the power has been restored? All of a sudden all the lights come back on, and there's a beep from the microwave letting you know that the clock needs to be reset.

The lights are an example of low-voltage release (LVR). The **switches** that control the lights were closed when the power went out, stayed closed during the power outage (the period of "low voltage"), and when the power was restored, the switches were all still in their "ON" position. The load was "released" after the period of "low-voltage."

Low-voltage release (LVR) is very useful for circuits where re-energization after a brief or temporary loss of power is safe and desirable. Some examples include lighting circuits, sump pumps, refrigeration circuits, and ventilation circuits. These are examples of circuits where their failure to re-energize after a power disruption could lead to damage of property (sump pump) or endangerment of safety (parkade ventilation).

The microwave was an example of low-voltage protection (LVP). The timer controlling the microwave shut off when the power went out, and when the power returned, the **control circuit** of the microwave waited for further input from a human operator.

Low-voltage protection (LVP) is desirable when the sudden activation of a machine or other electrical load could cause damage or injury. Some examples would include any rotating machinery (table-saws, lathes) or moving conveyor belts. These are examples of circuits where the sudden re-energization could surprise or injure a person working nearby.

There are many ways that circuits can be equipped with either low-voltage release (LVR) or low-voltage protection (LVP), but two of the simplest are with **two-wire** and **three-wire circuits** respectively.

As a rule, if a circuit uses a magnetic contactor and a holding contact, it will provide low-voltage protection (LVP).

If it uses maintained contacts, then it most likely provides low-voltage release (LVR).

Low voltage release (LVR)
Versus low voltage protection
(LVP)

A BCcampus element has been excluded from this version of the text. You can view it online here: https://pressbooks.bccampus.ca/basicmotorcontrol/?p=24 (https://pressbooks.bccampus.ca/basicmotorcontrol/?p=24#pb-interactive-content)

3.

Overload and Overcurrent Protection

Click play on the following audio player to listen along as you read this section.

A BCcampus element has been excluded from this version of the text. You can view it online here: https://pressbooks.bccampus.ca/basicmotorcontrol/?p=26 (https://pressbooks.bccampus.ca/basicmotorcontrol/?p=26#pb-interactive-content)

Inrush Current

When a motor is first started, before the shaft has a chance to pick up speed and begin to rotate, the characteristics of the stator coil are that of a short circuit. As such the motor starts to draw very high values of **current**. This current creates a magnetic field that causes the motor shaft to spin, and that spinning action creates a counter-EMF (CEMF), which limits the current to its normal running value.

The initial high value of current is called **inrush** and can cause severe line disturbances and nuisance tripping if **fuses** and **circuit breakers** are not sized accordingly.

Overload

The term "**overload**" describes a moderate and gradual rise in the value of current over a relatively long period of time. It is caused by excessive amounts of current drawn by a motor, which may be as high as six times the rated current. This is caused by too much load on a motor. Systems are protected by **overload protection relays**. While overloads are allowed for a short time (usually minutes), prolonged overloads will use thermal action to cause a protective device to trip.

Overcurrent

The term "**overcurrent**" (sometimes called a short circuit or a ground fault) describes a sharp and fast rise in current over a short period of time (fractions of a second). Circuits and equipment are protected from overcurrent situations by fuses or circuit breakers.

In these cases, the value of current is far greater than the nominal line current and can indeed be anywhere from six times to many hundreds of times greater the normal rated current value of current.

There are several causes of overcurrent situations. For example, when a bolted fault occurs—either a line to ground or a line to line fault. This causes a very large value of current to be drawn because of the inversely proportional relationship between the **resistance** of a circuit and the current drawn.

Another less intuitive cause of short circuits is when an induction motor starts. When a three-phase induction motor is first energized, the stator windings consist of a very low resistance path. This draws a very large inrush current which is indistinguishable from a standard short circuit, except that it quickly drops down to the rated value of current drawn by the motor. This is due to the CEMF (counter-electromotive force) developed by the rotating shaft of the motor. When the motor is spinning, a CEMF limits the current to safe values. When the motor is not spinning, a very large value of current is drawn from the source. This current is sometimes called **locked-rotor current**, and **motor starters** and overcurrent devices must be rated to safely handle this value of current.

Effects of short circuits

Two of the main negative outputs of overcurrents are:

- **Thermal energy**: High values of current will create lots of heat, which can damage equipment and wires. Thermal energy can be expressed by I^2t (current squared times time)—the longer the fault persists, the greater the potential thermal damage.
- **Mechanical forces**: High-fault currents can create powerful magnetic fields and exert huge magnetic stress on busbars and equipment, sometimes warping them out of shape and creating other problems.

Large values of fault current can cause damage very quickly, so overcurrent protective devices must act very quickly to clear the fault. There are two main categories of overcurrent protective devices: fuses and circuit breakers.

Fuses

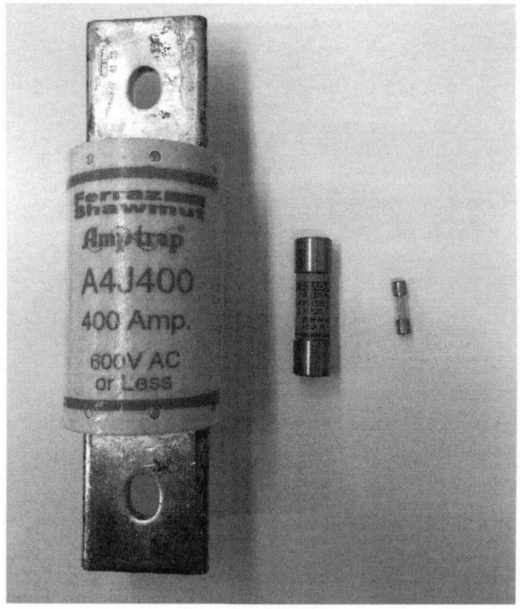

Fuses

A **fuse** is a simple device that protects the conductors and equipment of a circuit from damage due to higher than normal fault values. It is designed to be the weakest link in a circuit.

A fuse is an insulated tube containing a strip of conductive metal (fuse-link) that has a lower melting point than either copper or aluminum. The fuse link has narrow, resistive segments that concentrate the current and cause the temperature at those points to rise.

In a short circuit, the fuse elements burn open in just a fraction of a second. The higher the values of fault current, the faster the fuse will react.

In an overload situation, the fuse elements can take many seconds or even minutes before thermal actions cause the fuse link to melt open.

Fuses come in two categories: Fast-acting fuse (Type P) and time-delay fuse (Type D).

Fuses used in motor circuits have to withstand the intense inrush current when the motor is started, and so we use time-delay fuses, also known as “dual-element fuses.”

Common ratings

All overcurrent devices must be operated within their rated values. Three of the most important ratings are voltage, current, and interrupting capacity.

Voltage rating

Fuses and circuit breakers must be rated for at least the value of the **voltage** of the circuit they are designed to protect.

When a fuse or circuit breaker interrupts a fault current, it must safely extinguish the arc and prevent it from reestablishing itself. Therefore, the **voltage rating** of a fuse or circuit breaker must be equal to or greater than the system voltage.

For example, a fuse rated at 240V RMS will be acceptable for use in a 120V circuit. However, it would exceed the fuse's voltage rating to use it in a 600V circuit.

Continuous-duty rating

Continuous-duty rating describes the maximum rated value of RMS current that the overcurrent device is designed to handle on a continuous basis without tripping. Generally speaking, the **ampere** rating of the fuse or breaker should not exceed the current carrying capacity of the circuit, but there are exceptions, such as certain motor circuits.

Interrupting capacity

When a short circuit or ground fault occurs, the circuit resistance drops to effectively zero **ohms**, causing very large values of current to flow. This extremely fast rise in fault current can cause damage to wires and equipment through overheating and must be extinguished as quickly as possible.

The **interrupting-capacity (IC) rating** of an overcurrent device is the maximum fault current that the device can interrupt without damage to itself. Most circuit breakers and fuses have an IC rating of 10,000 amps.

For systems capable of larger fault currents, high-rupture capacity (HRC) fuses can interrupt currents up to 200,000 amps by using an arc-quenching filler such as silica sand to help interrupt the fault.

4.

Single-Phase Systems vs. Three-Phase Systems

Click play on the following audio player to listen along as you read this section.

A BCcampus element has been excluded from this version of the text. You can view it online here: https://pressbooks.bccampus.ca/basicmotorcontrol/?p=133 (https://pressbooks.bccampus.ca/basicmotorcontrol/?p=133#pb-interactive-content)

In electrical systems, we use the terms "single-phase" and "three-phase" fairly often, so a brief description of them will help us moving forward.

Single-phase systems are the simplest electrical circuits. They require only two lines: one for power to go in and the other is a return path for current. These are often called Line 1 and Line 2, or Line 1 and Neutral. Current only has one path to travel in a single-phase circuit, and all of the **control circuits** that we will be looking at are single-phase.

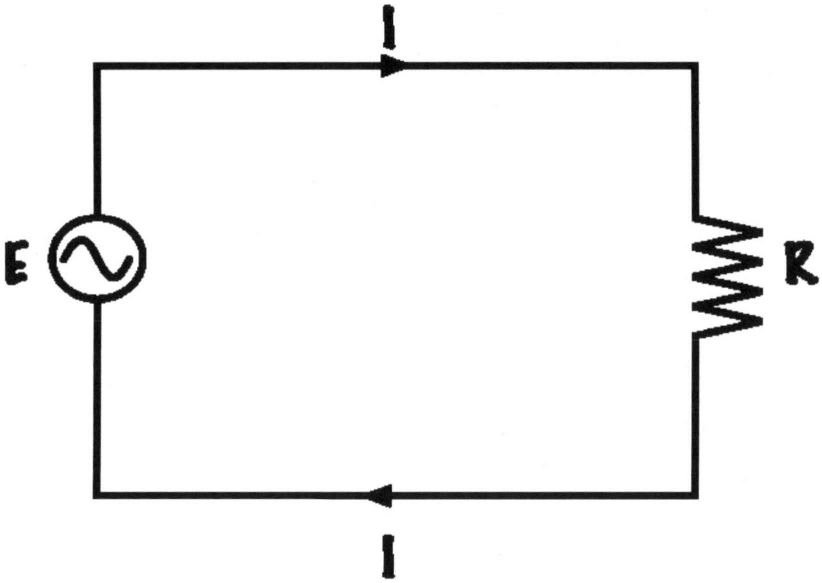

AC single phase circuit

Three-phase systems are bit more complex. They use three current carrying conductors, called Line 1,

Line 2, and Line 3, which have a 120° phase shift in the **voltage** and current waveforms between them. Each of these conductors are connected to a three-phase load, like a three-phase motor.

When in operation, a balanced three-phase load (such as a motor) has each of its three line's current values cancel each other out, and so it does not require a return conductor. These loads can be connected in Wye or Delta configuration, which will be covered later in this text.

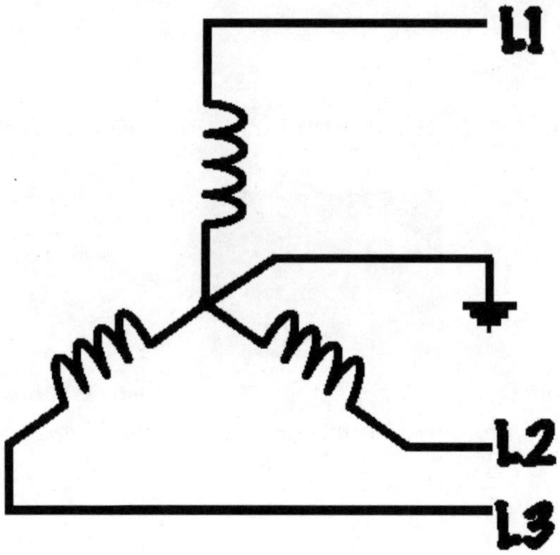

3-phase circuit

Unbalanced three-phase loads are mainly connected in the Wye configuration where the central point is used as a neutral to carry any stray return currents. In practice a motor is always a balanced three-phase load.

All of the **power circuits** that we will be looking at will be three-phase.

All of the control circuits that we will be looking at will be single-phase.

5.

Plugging vs. Anti-Plugging

Click play on the following audio player to listen along as you read this section.

A BCcampus element has been excluded from this version of the text. You can view it online here: https://pressbooks.bccampus.ca/basicmotorcontrol/?p=173 (https://pressbooks.bccampus.ca/basicmotorcontrol/?p=173#pb-interactive-content)

When a three-phase motor is first energized, it experiences an **inrush current** and harsh starting torque. If, for example, a motor is spinning in the forward direction and is stopped and suddenly reenergized in the reverse direction before the shaft of the motor has time to come to a complete stop, the motor will be said to have been "plugged."

Plugging is hard on motors, and if they are directly attached to high inertia loads, the shaft of the motor can shatter. Plugging also causes excess heating of the conductors that make up the coils inside the motor, as well as the power contacts of the **motor starter**. If a motor is expected to make sudden reversals of direction, then the motor starters should be increased in size and horsepower rating to accommodate for this more severe operation.

Some circuits use this plugging action to quickly decelerate and stop a motor. For example, in certain machine-tool operations where instantly stopping motor drives may be necessary. We will be examining a **plugging circuit** in greater detail further in this text.

If the sudden reversal of a motor is considered undesirable, then our **control circuit** can be designed to provide **anti-plugging**, which can be provided by either a **normally closed zero-speed switch** or off-delay timer relays.

As a general rule, the control scheme for plugging includes a reversing magnetic starter and a rotary-motion switch called a zero-speed switch. The switch assembly is usually connected directly to the shaft of the motor or the driven equipment so that when the motor spins in a particular direction, the **contacts** change their state (normally closed contacts open, and normally open contacts close).

Normally open contacts provide plugging while normally closed contacts provide anti-plugging.

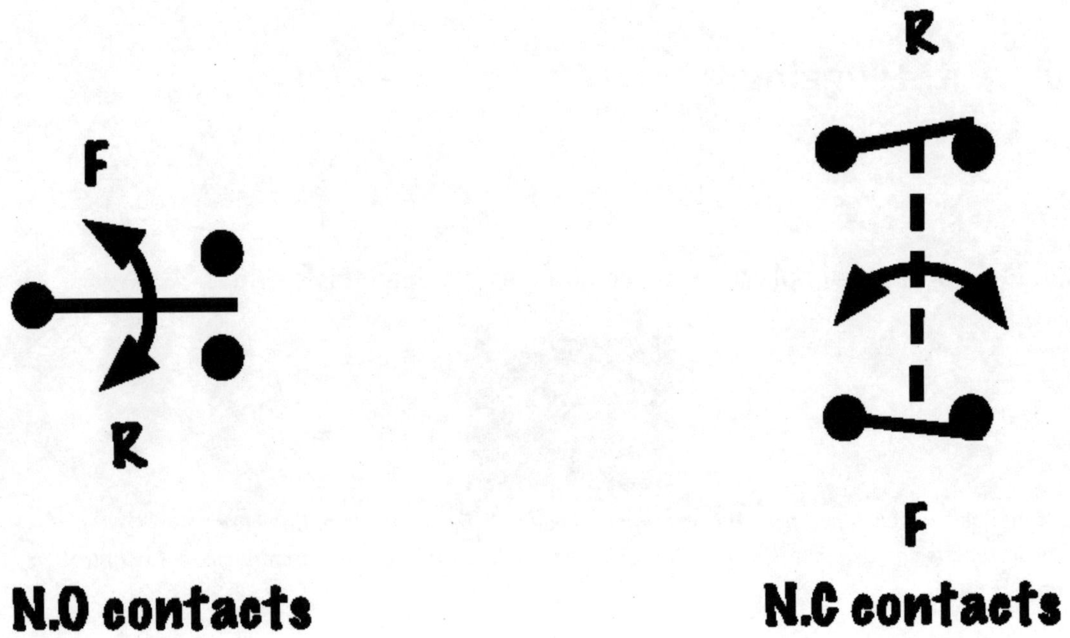

Zero-speed switch contacts

6.

Self Test 1

An interactive or media element has been excluded from this version of the text. You can view it online here: https://pressbooks.bccampus.ca/basicmotorcontrol/?p=513 (https://pressbooks.bccampus.ca/basicmotorcontrol/?p=513#pb-interactive-content)

If using the print, PDF, or eBook copy of this book, navigate to the above link to complete the quiz. However, the quiz questions are also provided at the end of the book for offline use: Offline Copies of Chapter Quizzes

II

Pilot Devices

Learning Objectives

After studying this section, students should be able to:

- Describe common control devices such as momentary-contact pushbuttons.
- Describe the behaviour of common switches, including their terminology, uses, number of switchable poles, and throw positions.
- Be familiar with basic pilot devices such as liquid-level switches, pressure switches, limit switches, etc.
- Describe the action and purpose of the zero-speed switch.

7.

Switches and Pushbuttons

Click play on the following audio player to listen along as you read this section.

A BCcampus element has been excluded from this version of the text. You can view it online here: https://pressbooks.bccampus.ca/basicmotorcontrol/?p=31 (https://pressbooks.bccampus.ca/basicmotorcontrol/?p=31#pb-interactive-content)

Pilot Devices Overview

These are some of the most basic control devices that we use to start and stop our motors. They can be provided with **fuses** for **overcurrent** protection, and if a **switch** is horsepower rated, it is designed to make and break line current to a motor.

Pushbuttons

Pushbuttons are momentary **contacts** that come available as either **normally open (NO)** or **normally closed (NC) contacts**, although many pushbuttons come with both NO and NC sets of contacts.

When you depress the button, you change the state of the contacts, and when you release the button, a spring returns the contacts to their original state.

These are ideal for use in low-voltage **control circuits**.

Pushbuttons

Switches

Switches are identified by the number of conductors (poles) they connect to, and the number of positions (throws) they can switch to. Switches are also rated for **voltage** and **power** and must be operated within their limits.

Single-pole, single-throw (SPST)

This switch is used where we only need to break a single line in a 120V single-phase supply, providing that it is connected to an ungrounded-circuit conductor.

Single-pole, double-throw (SPDT)

This switch connects a single-line conductor to either of two possible switch legs. A common household example of this is in a three-way switch loop, which allows control of a single load (usually a light) from two different locations.

In industrial settings, it is commonly used as a Hand – Off – Auto selector switch.

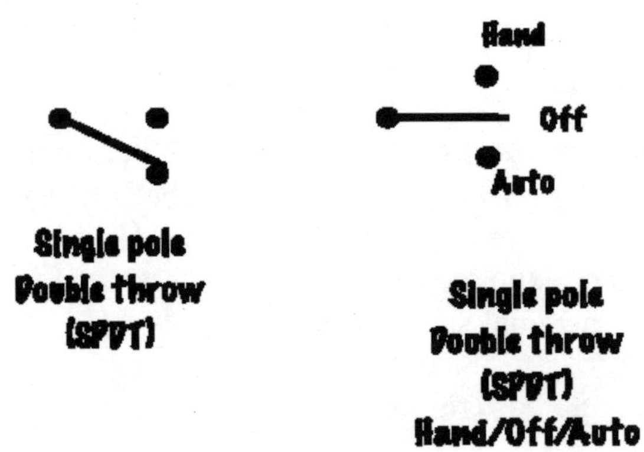

Single-pole, double-throw

Double-pole, double-throw (DPDT)

The most common application of these devices is as a 4-way switch, which is used in conjecture with two three-way switches to allow for control from three or more locations. As a general rule, a light can be controlled from any number of locations, providing you start and end your switching with three-way switches and use as many 4-way switches in between as necessary.

Double-pole, double-throw

Double-pole, single-throw (DPST)

This switch looks like a standard SPST switch from the outside but inside, instead of interrupting only one current-carrying conductor, it interrupts two. This makes it suitable for controlling and isolating loads that operate at 240V single-phase.

Triple-pole, single-throw (3PST)

These are used to interrupt current to **three-phase circuits** and motors. With a single external handle, three current-carrying conductors can be opened or closed at once. The switches often come with housing for the mounting of fuses to provide overcurrent protection. The switch must be horsepower rated if it is used to interrupt current to a motor.

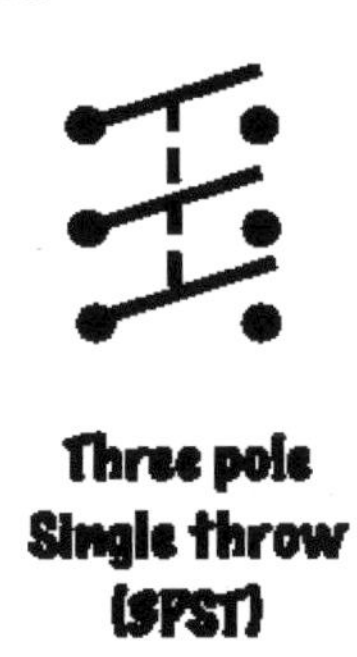

Isolating switch

An **isolating switch** is connected upstream of a **power circuit** and is NOT horsepower rated and so it is not meant to interrupt current flow. Isolating switches are not meant to control motor loads. Rather once a motor has been properly shut off, an isolating switch can be used for lockout purposes. An isolating switch can have any number of pole contacts, but to be used for lockout purposes it must provide only a single throw option.

8.

Common Pilot Devices

Click play on the following audio player to listen along as you read this section.

A BCcampus element has been excluded from this version of the text. You can view it online here: https://pressbooks.bccampus.ca/basicmotorcontrol/?p=33 (https://pressbooks.bccampus.ca/basicmotorcontrol/?p=33#pb-interactive-content)

Switches, regardless of their number of **contacts** and throw positions, can be controlled in one of two ways. The first and most obvious is manually, whereby a person or operator physically moves the operating handle of the switch, changing the state of its contacts. The second category is to automatically control the switch via some sort of **process variable**.

Many processes require automatic switching, independent of human operators for safety, accuracy, and efficiency. For instance, motors are often controlled by what are known as **pilot devices** that sense external conditions and provide two-wire control.

There are nearly limitless process variables which can be monitored by pilot devices, such as liquid level, proximity, temperature, and pressure.

All pilot devices share the same two-step process:

1. Sense external input.
2. Change the state of electrical contacts.

This text will focus mainly on the second of these two stages, as the connection of electrical contacts into our **control circuits** is where we will see the most utility for our pilot devices. Indeed an entire textbook could be written about the many types and configurations of various pilot devices, but this is beyond the scope of this book. Instead, we will examine simple control devices that respond to the condition of a process variable such as temperature, pressure, liquid level, flow, position or proximity by activating a set of contacts.

Float Switches

Float switches, also known as liquid-level switches, are pilot devices used primarily in pump-motor applications to respond to changes in liquid level. There are many different types of float switches, but the simplest mechanical arrangement is a pivoted arm that has the electrical contacts fastened at one end and a rod connected to a float at the other. As the water level rises or falls, the float moves with the water level and the contacts change their state accordingly.

For typical tank-filling services, float switches close their electrical contacts at a low-liquid level and open at a high level, while for sump pump purposes, the float switch closes the circuit at a high-liquid level and opens the circuit at a low-liquid level.

Temperature Switches

Temperature switches are used to provide automatic control in heating and cooling systems. Simple thermostats often use a coiled bimetallic element to activate a switch contact. This is sufficient for the simple heating and cooling of rooms.

For industrial processes that require monitoring of material temperature, a remote-bulb temperature switch is often used. The bulb is placed inside the medium to be measured and is linked via a capillary tube to a bellows. As the temperature of the fluid in the bulb rises, its volume expands and causes an internal pressure change. This change in pressure results in the movement of the bellows, activating a set of switch contacts.

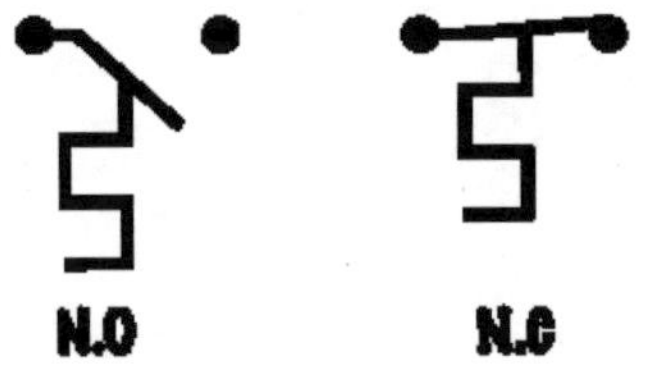

Pressure Switches

A pressure switch is a device that is used to sense a rise or fall in pressure and is usually pipe-connected to the vessel containing the medium (such as water, steam, air, or gas) that operates the switch. Pressure switches are commonly used as safety devices to protect system equipment from either dangerously high pressure or low pressure.

When dealing with pressure controls, the term “cut in” refers to the pressure of the system when the switch closes, and this normally applies to decreasing pressure. The term “cut out” refers to the pressure of the system when the switch opens and normally applies to increasing pressure.

While the range of a pressure switch describes the overall pressure over which the switch can operate, the sensitivity of its contacts can usually be adjusted and the differential describes the difference between the cut-in and the cut-out pressure of the switch.

Flow Switches

A flow switch is activated by the movement of either liquid or air through a pipe or duct. For example, to ensure that heating elements are not energized if air or water is not flowing across heat exchangers, a flow switch might be utilized.

A simple type of flow switch uses a pivoted arm with contacts at one end and a paddle on the other, which is inserted into a pipe so that the flow of fluid activates a set of electrical contacts.

Mechanical-Limit Switches

Mechanical-limit switches are commonly used to prevent over-travel in elevators and along conveyor belts. The operation of a limit switch relies on an external force striking an operating arm or lever, which in turn activates a set of control contacts. Once the force is removed, the operating arm returns to its normal position and the contacts reset.

Special types of limit switches can provide a maintained-contact action, whereby the contacts remain in their changed position even after the initial force is removed from the activation arm. These types of limit switches need to be reset between uses, usually by simply moving the operating mechanism in the opposite direction.

Limit switch contacts are frequently used in applications where **normally open contacts** are mechanically held closed, or **normally closed contacts** are mechanically held open. Control schematics usually display symbols to indicate the state of the switch contacts when the system is in the "off" condition. For example, the symbol for a "normally open held closed" limit switch indicates that the contacts have been wired normally open, but when the circuit is put into its normal off state, part of the machine keeps the contact held closed.

Other types of pilot devices, such as those used in pressure and flow switches, can be configured in a similar way, but it is seen most often with mechanical-limit switches.

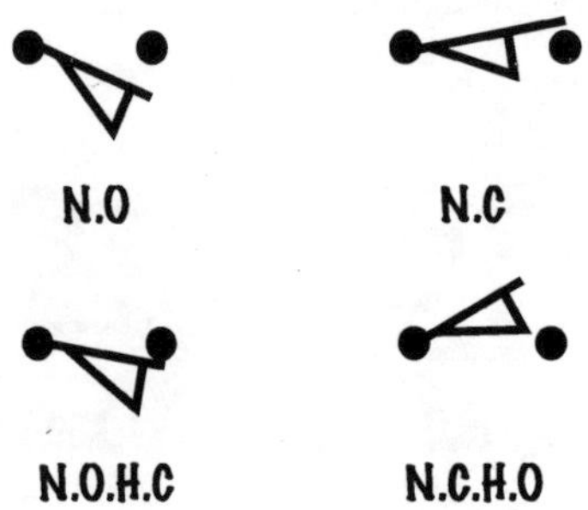

Proximity Switches

A proximity switch is a type of limit switch that does not require physical contact in order to function but rather senses the presence of particular materials. There are three main styles of proximity switches, each relying on different physical phenomenon to operate.

Inductive-proximity switch

Inductive proximity switches detect the presence of both ferrous and non-ferrous metals (ferrous refers to materials containing iron). They rely on the creation and monitoring of an electromagnetic field. An example is traffic sensors embedded into the pavement that sense the presence of the metallic body of the car and control the signal action of the traffic lights.

Capacitive-proximity switch

Capacitive-proximity switches detect the presence of almost any material. They rely on the creation and detection of an electrostatic field. Examples include the interactive touch screen of almost every modern electronic device, including smart phones and tablets.

Both inductive and capacitive types of proximity sensors use an oscillator circuit to generate a sensing field and series of solid-state devices to sense any change in the field, which in turn activates a set of external contacts.

Photoelectric switch

Photoelectric switches are proximity switches which emit and detect a beam of light from an infrared

LED. If the beam is interrupted or interfered with, this is sensed by the receiver and a signal is sent to another series of solid-state devices, which in turn then operate a set of contacts. Examples include the sensors on automatic doors, including elevator doors, as well as touchless faucets that activate whenever a person places their hands in the sink to be washed.

9.

Zero-Speed Switch

Click play on the following audio player to listen along as you read this section.

A BCcampus element has been excluded from this version of the text. You can view it online here: https://pressbooks.bccampus.ca/basicmotorcontrol/?p=50 (https://pressbooks.bccampus.ca/basicmotorcontrol/?p=50#pb-interactive-content)

A **zero-speed switch**, also called a plugging switch, is commonly connected, either directly or via a belt, to the shaft of the motor or the equipment that it drives and consists of a set of **contacts** that are activated by centrifugal force from the rotation of the shaft mechanism. The sets of contacts can come as either **normally open** or **normally closed**. Each set will change their state when subjected to shaft rotation.

Several symbols are used for zero-speed switch contacts, and the curved arrow represents the rotation of the motor to either open or close the contacts depending on their normal state.

The threshold of when these contacts open or close can be easily adjusted with a screw through the switch housing.

Since the slightest rotation can activate the switch contacts and introduce a hazard, some plugging switches are equipped with a special solenoid coil called a **lockout relay**. The lockout relay incorporates a mechanism to disable the contacts until the motor itself has been energized.

Zero-speed switches are used in circuits to provide either **plugging** action (momentarily reversing a motor to stop it) or **anti-plugging** action (preventing the reversal of a motor until it has come to a rest).

As a rule, a zero-speed switch with normally open contacts will provide plugging action, and a zero-speed switch with normally closed contacts will provide anti-plugging action.

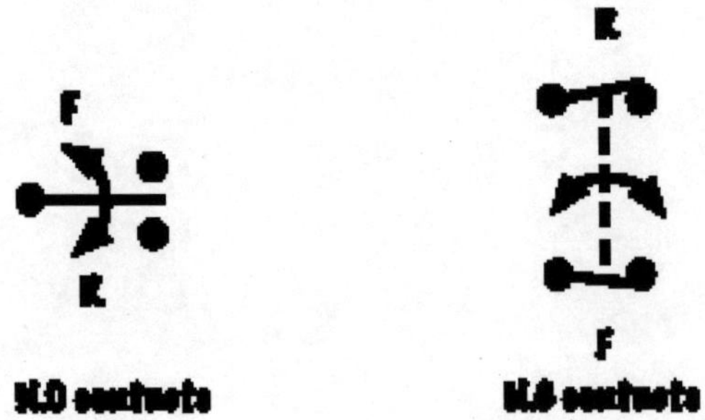

Zero-speed switch contacts.

10.

Self Test 2

An interactive or media element has been excluded from this version of the text. You can view it online here: https://pressbooks.bccampus.ca/basicmotorcontrol/?p=515 (https://pressbooks.bccampus.ca/basicmotorcontrol/?p=515#pb-interactive-content)

If using the print, PDF, or eBook copy of this book, navigate to the above link to complete the quiz. However, the quiz questions are also provided at the end of the book for offline use: Offline Copies of Chapter Quizzes

III

Motor Starters and Contactors

Learning Objectives

After studying this section, students should be able to:

- Describe common electrical ratings as they apply to motors and motor starters.
- Describe and identify common styles of overload-protection devices.
- Describe and identify manual motor starters and their applications.
- Describe and identify the components and functions of magnetic motor starters used for both single direction and forward/reverse applications.
- Describe the function and application of control relays.
- Describe the function and application of timer relays including on-delay relays and off-delay relays.

11.

Overview

Click play on the following audio player to listen along as you read this section.

A BCcampus element has been excluded from this version of the text. You can view it online here: https://pressbooks.bccampus.ca/basicmotorcontrol/?p=38 (https://pressbooks.bccampus.ca/basicmotorcontrol/?p=38#pb-interactive-content)

A **motor starter** is a device that is designed to safely start and stop a motor, and provide running (**overload**) protection.

A starter can be a simple two-wire manually operated **switch** or a more complex magnetically-operated forward/reverse motor starter. Regardless of its design, all motor starters must be used within their **voltage** and **power** ratings and must provide overload protection. One of the ways that motor starters provide overload protection is through the use of overload relays (OLR).

Common Ratings

There are three significant electrical ratings for motor starters: voltage, horsepower, and ampacity (current value).

The voltage rating of a motor starter must be rated for at least the operating voltage of the motor that it controls. This rating is specified on the motor and starter nameplates.

The ampacity rating of a starter usually refers to the overload protection device. The selection of the proper overload heater depends on the FLA (full-load-amps) current drawn by the motor. FLA is usually specified on the motor's nameplate and most manufacturers supply a table for heater-size selection.

Ambient temperature can also have an effect on overload heater sensitivity. Precautions should be taken when operating motor starters in very high or low ambient temperature environments.

The horsepower rating of a starter must meet or exceed the horsepower rating of the motor it controls at its listed operating voltage.

Horsepower is always given at a specified voltage. For example, a starter rated 3 horsepower at 240 V must not be used to control a 3 horsepower, 120 V. Instead it would most likely be rated for only 1.5 horsepower at 120 V.

12.

Overload Relays

Click play on the following audio player to listen along as you read this section.

A BCcampus element has been excluded from this version of the text. You can view it online here: https://pressbooks.bccampus.ca/basicmotorcontrol/?p=125 (https://pressbooks.bccampus.ca/basicmotorcontrol/?p=125#pb-interactive-content)

An **overload relay** has two basic parts:

1. A heater element that is connected in **series** with the power line to the motor. All current drawn by the motor must pass through the heater element.
2. A set of **normally closed contacts** that are connected in series with either the lines feeding the motor (manual starters) or the coil of the magnetic contactor (magnetic starters). The types of relays most commonly seen are the bimetallic strip and the melting-solder pot assembly.

Bimetallic Strip

The bimetallic strip consists of two dissimilar metals with different heating coefficients. As they heat, they expand at different rates, which causes them to bend or deform at a preset temperature. This bending action can open or close a set of **contacts**.

When used in an overload device, the bimetallic strip is mechanically linked to a set of normally closed electrical contacts. When an overload occurs, the bending action opens the set of normally closed contacts, interrupting the current to the circuit.

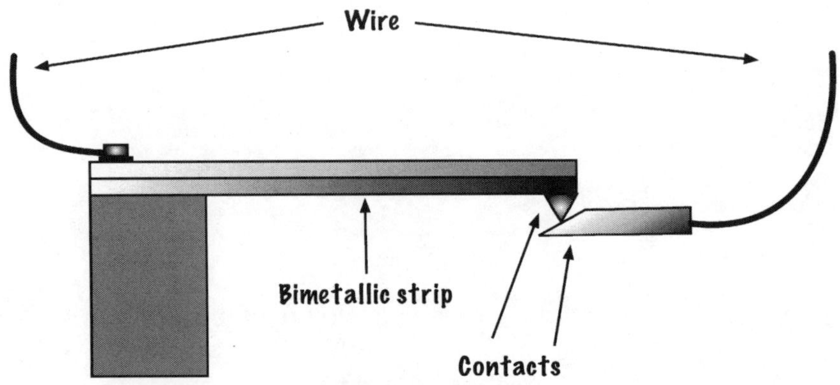

Bimetallic contact in a normally closed position

Here the normally closed contacts are allowing current to pass through them, while the heat source is beginning to deform the metal.

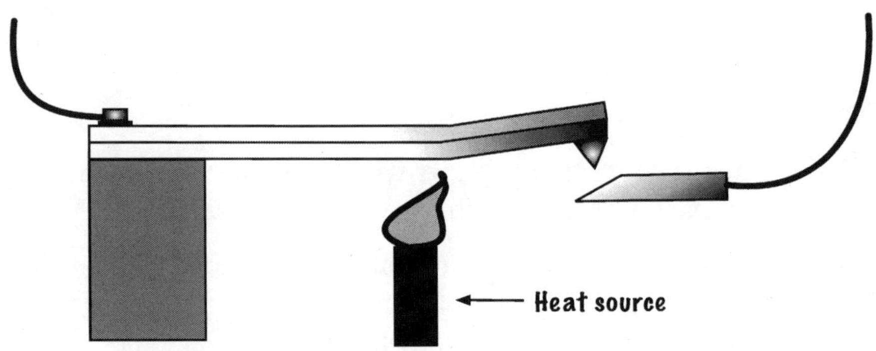

Bimetallic strip in open position

The heat source has caused the grey-shaded metal (the piece on the bottom) to expand faster than the blue-shaded metal (the piece on the top) and so has opened the set of normally closed contacts, thus interrupting the current flow to the motor.

Melting Solder Pot

Solder pot closed contacts

The melting solder pot consists of a heater element, a solder-pot assembly, ratchet wheel, and a set of normally closed contacts.

A spring is held under tension by the ratchet wheel. If the wheel is allowed to spin, then the spring will push upwards and open the set of normally closed contacts. The wheel is held in place by the solder inside the solder-pot assembly. Different levels of tin and zinc in the solder change the melting temperature, allowing for use at many different current ratings and ambient temperature settings.

If an overload current is sensed by the heater elements for too long a time, then the alloy becomes a liquid, allowing the spring to push open the normally closed contacts. This causes the line contacts to trip open and interrupt current flow to the motor.

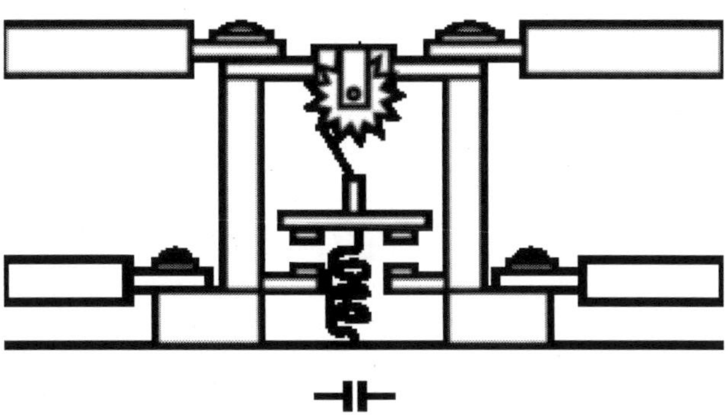

Solder pot open contacts

Both the bimetallic strip and melting solder-pot rely on thermal energy to trip their elements. As such, a cool-down period is required before the contacts can be reset. Once the relay cools, the bimetallic strip will return to its normal position, or the melted solder will solidify and the ratchet wheel can be reset to close the line contacts again.

Other Types of Overload Relay

Some modern motor control systems incorporate real-time current transformer monitoring applications that use integrated computer **control circuits** to protect against motor overloads. These systems can be interconnected with networked PLC's and other safety equipment.

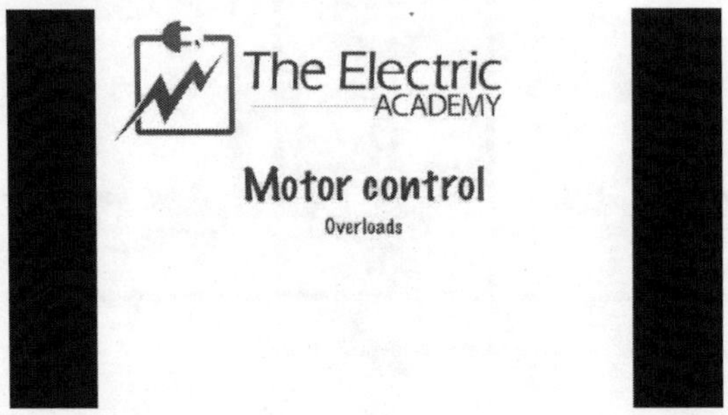

A BCcampus element has been excluded from this version of the text. You can view it online here: https://pressbooks.bccampus.ca/basicmotorcontrol/?p=125 (https://pressbooks.bccampus.ca/basicmotorcontrol/?p=125#pb-interactive-content)

13.

Manual Motor Starters

Toggle

Small, fractional-horsepower motors are often controlled using a simple toggle **switch**. Most toggle-switch starters are rated at maximum 1 horsepower at 230V. They may be for single-pole or double-pole switching, and the ON and OFF positions are clearly marked on the lever.

The switch is similar in appearance to a standard light switch, but the main difference is that the **motor starter** has **horsepower-rated contacts**, which means that it can safely make and break the current drawn by the motor.

It also contains a thermal-overload device to provide overload protection by sensing any higher-than normal currents drawn by the motor. If a motor **overload** occurs, the excess current drawn by the motor causes the heater element to heat up, which automatically trips the switch contacts and breaks the circuit. The handle moves to the middle or tripped position, and after a cooling period, the toggle switch can be reset by pushing the lever into the fully off position and then switching it on again.

The compact construction of this manual-starter switch allows it to be installed directly on driven machinery and in tight spaces.

Pushbutton

Pushbutton motor starters are used for single-phase motors that are typically rated at a maximum 5 horsepower at 230V. They are similar in operating principle to toggle switches, except that they use two separate buttons to activate contacts. To start the motor, depress the start button so that it moves inward. A **mechanical interlocking device** forces the stop button outward. Once activated, these buttons remain in position until the opposite button is pressed.

If an overload occurs, the contacts can be reset by pushing firmly on the stop button after allowing for a cooling period.

Drum Switch

The drum switch is a rotary-action selector switch that is commonly used to manually control speed and to change the directions of motors. The handle in the centre at the top can be moved radially to various positions, turning a central shaft. The moving contacts on the shaft connect to a series of contacts on the perimeter of the drum.

When used to control single-phase motors, they are typically rated up to 3 horsepower at 240V or 1.5 horsepower at 120V.

The drum switch does not provide overload protection. Additional overload protection must be installed when using drum switches.

All manual starters have maintained contacts. If the power fails while the contacts are closed, when power is restored, the load will automatically re-energize. As a general rule, manual starters provide two-wire control and **low-voltage release (LVR)**.

14.

Magnetic-Motor Starters

For the control of three-phase motors, magnetic contactors are used to open and close the power contacts in line with the motor. This allows for separation of the **control circuit** from the **power circuit**, providing greater safety for the operator and ease and convenience of wiring for the installer. Magnetic contactors also provide **low-voltage protection (LVP)** in the case of a power outage.

Magnetic contactors must also have built-in overload protection if they are to be used to control motors. The most common controllers for three-phase motors are the across-the-line magnetic starter, meaning that the motor is started with full line voltage.

NEMA vs. IEC

The difference between NEMA and IEC contactors is in their certifications and ratings. NEMA (National Electrical Manufacturers Association) is recognized in North America.

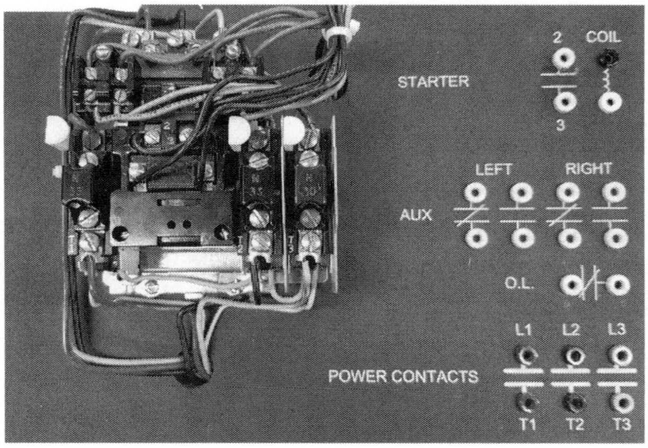

NEMA motor starter

IEC (International electrotechnical commission) is recognized in both North America and Europe.

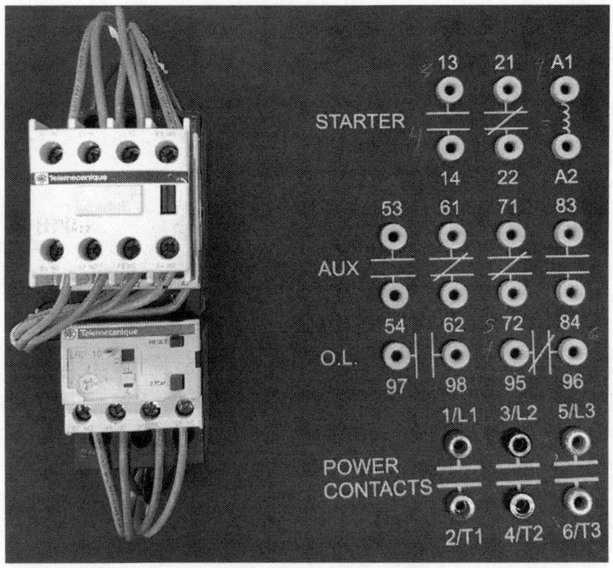

IEC motor starter with overload relay

As a general rule, NEMA equipment is more expensive and robust than IEC equipment, but IEC equipment is more versatile. And because is often cheaper, IEC equipment is more commonly seen in modern installations.

Contactor Construction and Operation

A magnetic **motor starter** has two basic parts: the magnetic contactor and an **overload relay**.

The magnetic contactor is a solenoid relay consisting of stationary contacts that are wired in **series** with the lines to the motor, an inductive coil wrapped around a magnetic core and a movable **armature** attached to moveable contacts. When an electric current is passed through the coil of wire, a magnetic field is created. This field, in turn, draws the armature towards it, causing the movable contacts to bridge the gap of the stationary contacts and thus energizing the motor. A spring is constantly trying to open the contacts but as long as there is a **voltage** present at the coil, magnetic forces will overcome the strength of this spring.

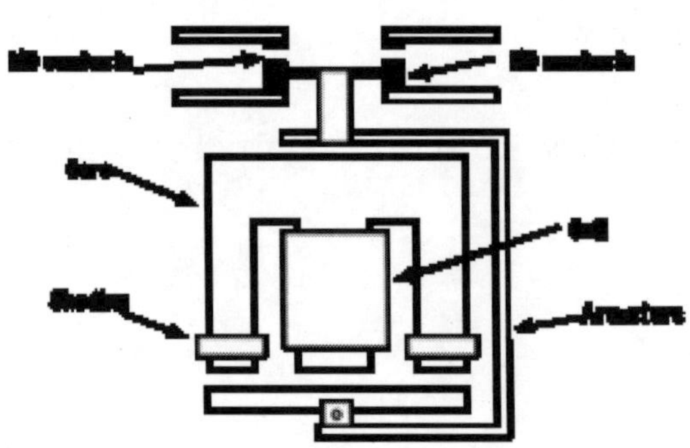

Contactor coil de-energized

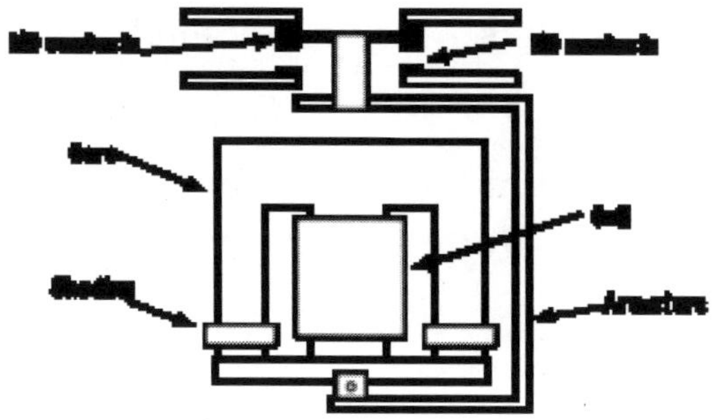

Contactor coil energized

When there is a power outage, however, and the current through the coil drops below a threshold, the spring will push open the contacts. If power is restored, the motor load will not re-energize but will instead require further input from the operator. This type of control is called three-wire control and it provides low-voltage protection (LVP).

For the control of three-phase motors, contactors are built with three sets of **horsepower-rated contacts**. Additional **auxiliary contacts** can be included as well. Relay contacts are usually silver-plated to improve their conductivity and although single-break contacts are used, most industrial quality relays use double-break contacts to improve their interrupting capability.

Coils are generally designed to activate at approximately 85% of their rated voltage and will not deactivate until the voltage falls below approximately 85% of the rated value. Usually a coil can withstand an overvoltage of up to 10% without damage to the coil.

> **Question:** If magnetic coils are power by AC, why don't their contacts open and close 120 times per second?
>
> **Answer:** Sometimes they do! If a magnetic contactor is making an unnatural "chattering" sound, it may be caused by a loose or faulty shading coil. Shading coils are simple closed loops of conductive material that, when exposed to the changing magnetic field of an AC circuit, creates its own magnetic field with a slight delay in period. This allows for a constant magnetic attraction between the moveable armature and the coil of the contactor. If a contactor is "chattering," it may need its shading coils repaired or replaced.

The overload relay (OLR) is similar in design to that used in manual motor starters. The key difference is that the **normally closed contacts** of the OLR are wired in series with the current that flows through the armature of the contactor coil. This ensures that if an **overload** occurs in any of the three power lines supplying the motor, the normally closed contacts of the OLR will open, and the contactor supplying power to the motor is disconnected from the circuit.

The key utility comes in separating the control circuit from the power circuit. Magnetic-motor starters, for example, can allow the control of a 50 horsepower 600 V three-phase motor (the power circuit) by simply energizing a 120V, 1A load.

This concept of motor starters as a load that controls other larger loads is key to our further understanding of basic motor control.

Combination Starters

A combination starter refers to a simplified modular device that contains three-phase disconnects, **overcurrent** protection, magnetic contactor and overload relays.

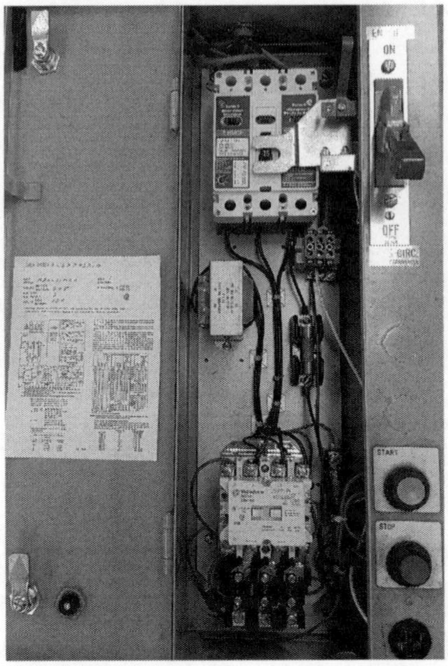

Combination starter

15.

Forward/Reverse Starters

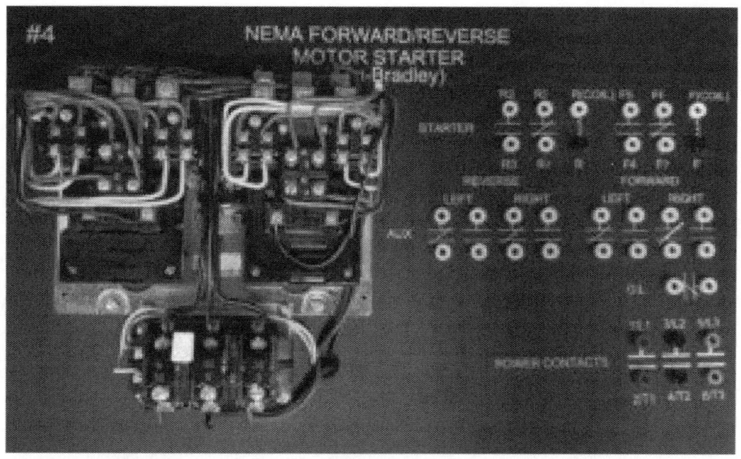

NEMA Forward Reverse motor starter with OLR

A forward/reverse magnetic **motor starter** is constructed from two regular contactors installed with a **mechanical interlock** that prevents both coils from pulling in simultaneously.

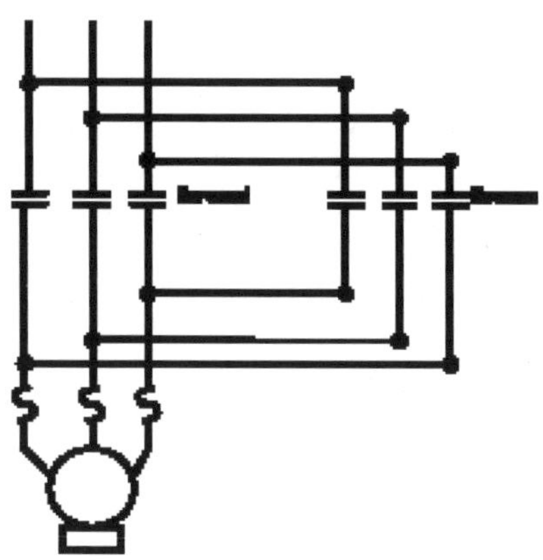
Forward/Reverse power circuit

They are connected so that under normal conditions, one coil will connect L1 with T1, L2 with T2, and L3 with T3. When the alternate coil is energized, Lines 1 and 3 are interchanged with Terminals 1 and 3, allowing for a phase shift in the 3-phase circuit, driving the motor and causing it to rotate in the opposite direction.

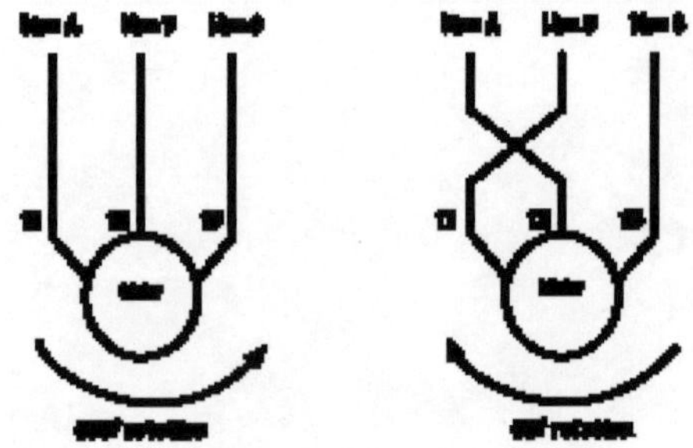

Reversal of three-phase motor rotation

Both coils power contacts feed through a single set of OLR to provide running protection in both directions.

To prevent unintentional energization of both the forward and reverse coil at the same time, the coils are equipped with **normally closed contacts**, which are then wired in **series** with the opposite direction's coil to act as an **electrical interlock**. This means that if the coil that drives the motor in the forward direction is energized, a normally closed set of contacts wired in series with the reverse coils will prevent that coil from being energized as long as the forward coil remains energized.

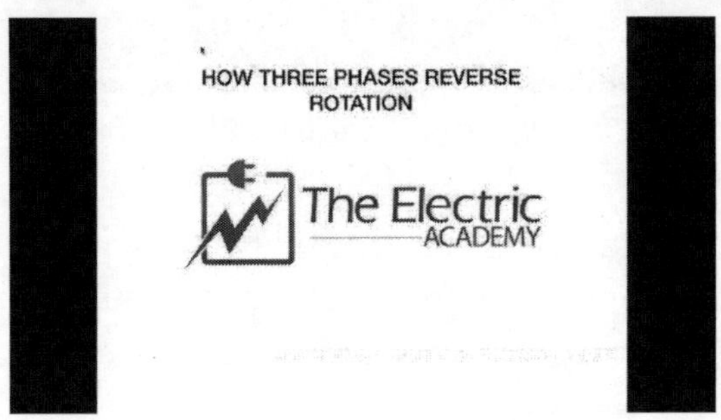

A BCcampus element has been excluded from this version of the text. You can view it online here: https://pressbooks.bccampus.ca/basicmotorcontrol/?p=44 (https://pressbooks.bccampus.ca/basicmotorcontrol/?p=44#pb-interactive-content)

16.

Control Relays

Control relays are magnetic devices that are used to remotely switch loads. They are functionally identical to a magnetic **motor starter**, save that control relays lack **horsepower-rated contacts** and built-in **overload** protection.

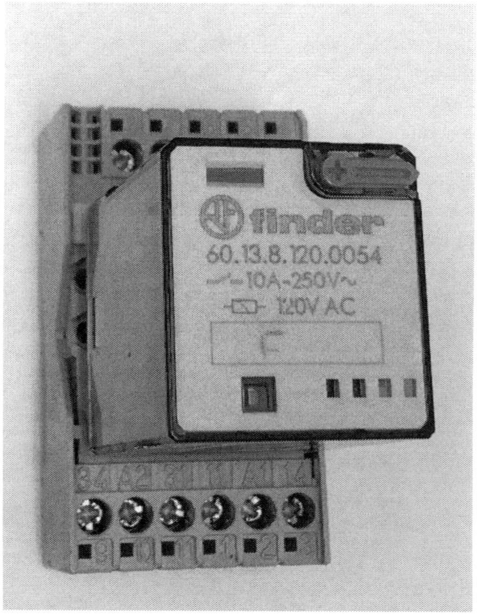

Control relay.

Since magnetic-motor starters are limited in the number of **auxiliary electrical contacts**, control relays are used where additional contacts are required to switch low-level current loads such as **pilot lights**, motor starters, or other relays.

Some control relays are the latching-type relay, in that after energizing the coil, the relay contacts remain mechanically latched in the closed position, even after power is removed from the coil. To open the relay contacts, the coil requires the momentary energization of a second coil to unlatch the coil.

The advantage of this type of relay is that it is not necessary to keep the coil continuously energized, which results in quieter operation and less current drawn from the source.

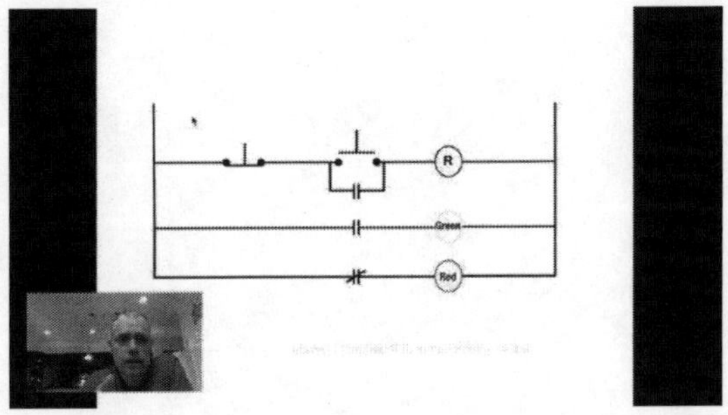

A BCcampus element has been excluded from this version of the text. You can view it online here: https://pressbooks.bccampus.ca/basicmotorcontrol/?p=46 (https://pressbooks.bccampus.ca/basicmotorcontrol/?p=46#pb-interactive-content)

17.

Timers

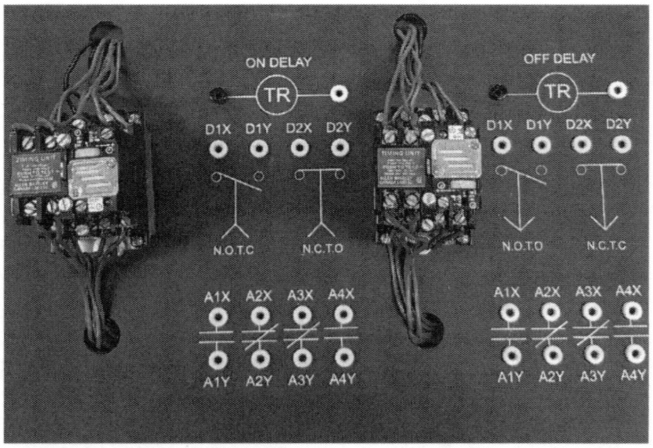

Pneumatic-timing relays.

Whenever a motor is started, it draws an initial **inrush current**. If several motors are to be started simultaneously, these combined inrush currents could cause line disturbances throughout the system and cause nuisance tripping of **overcurrent** devices.

If two or more motors are controlled from the same **pushbutton** station, it may be desirable to provide a time interval between their starting. One of the ways that we can accomplish this is to use timing relays.

Timers are similar in construction to **control relays** in the sense that they have a low-voltage **armature** and coil that opens or closes a set of **con**tacts when energized. The distinctive quality of timers is the addition of some sort of delay mechanism that prevents the **auxiliary contacts** from changing their state for some pre-set amount of time.

Timers come in two classes: On-delay timers and off-delay timers.

On-Delay Timers

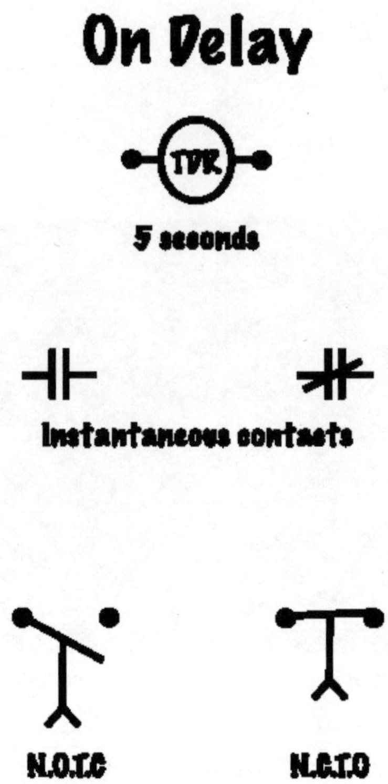

On-delay timer contacts

These are commonly used in circuits where we might want to start two motors from the same pushbutton station but with a slight delay to minimize line disturbances due to inrush. The timing contacts for the on-delay timer can be either **normally open, timed to close (NOTC)**, or **normally closed, timed to open (NCTO)**.

On-delay timers delay their action when the coil is energized and revert to their original state instantly when de-energized. Some timers will have instantaneous contacts in addition to their timed contacts.

When looking at the schematic symbol for timed contacts, the arrowhead indicates the direction of the timing action. Some **schematic drawings** show timed contacts as instantaneous contacts identified with their timing behavior.

Off-Delay Timers

Off-delay timer contacts

These are commonly used in circuits where we might want to stop two motors from the same pushbutton station but allow the second motor to run for an additional period of time. This arrangement is commonly used in furnace HVAC systems and conveyor belt circuits. The timing contacts for the off-delay timer can be either normally open, timed to open (NOTO), or normally closed, timed to close (NCTC).

Off-delay timers change their original state instantly when energized. And delay their action when the coil is de-energized. Some timers will have instantaneous contacts in addition to their timed contacts.

When looking at the schematic symbol for timed contacts the arrowhead indicates the direction of the timing action.

Some schematic drawings show timed contacts as instantaneous contacts identified with their timing behavior.

18.

Self Test 3

An interactive or media element has been excluded from this version of the text. You can view it online here: https://pressbooks.bccampus.ca/basicmotorcontrol/?p=517 (https://pressbooks.bccampus.ca/basicmotorcontrol/?p=517#pb-interactive-content)

If using the print, PDF, or eBook copy of this book, navigate to the above link to complete the quiz. However, the quiz questions are also provided at the end of the book for offline use: Offline Copies of Chapter Quizzes

IV

Diagrams

Learning Objectives

After studying this section, students should be able to:

- Describe common methods of drawing power and control circuits.
- Describe the different applications of schematic and wiring diagrams.
- Apply a numbering system to convert from one style of diagram to another.
- Identify and connect three-phase nine lead Wye and Delta wound motors.

19.

Schematic vs. Wiring Diagrams

One of the most frequently used diagrams in motor control work is the ladder diagram, also known as a **schematic diagram**. This diagrams uses symbols to identify components and interconnecting lines to display the electrical continuity of a circuit.

Ladder diagrams show how a circuit works logically and electrically. When troubleshooting or designing **control circuits**, we use "ladder" or "schematic" diagrams to represent how the circuit works but these diagrams do not show how equipment is physically laid out, nor do they represent how things are wired "in the real world." Two components might be right next to each other on the ladder diagram and 50 metres apart in the real world.

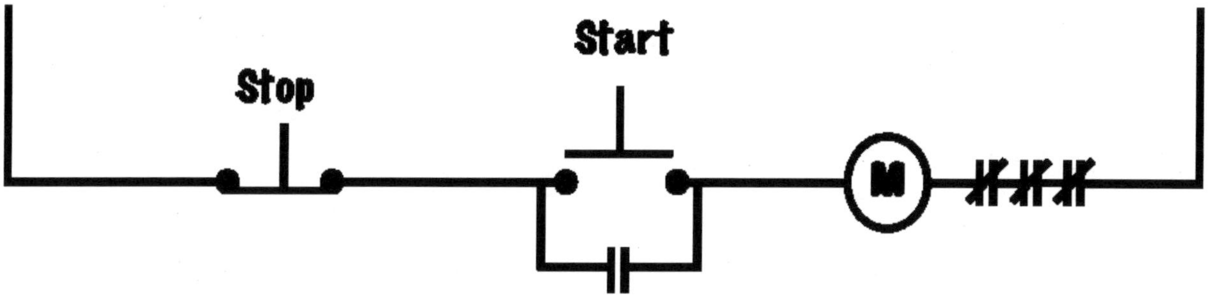

Schematic drawing

Wiring diagrams, also called connection diagrams, however, do show how equipment is laid out and the connections between them. A wiring diagram shows the relative layout of the components and the wire connections between them. This type of diagram shows the physical relation of all devices in the system, the conductor terminations between these devices, and are commonly used in motor control installations.

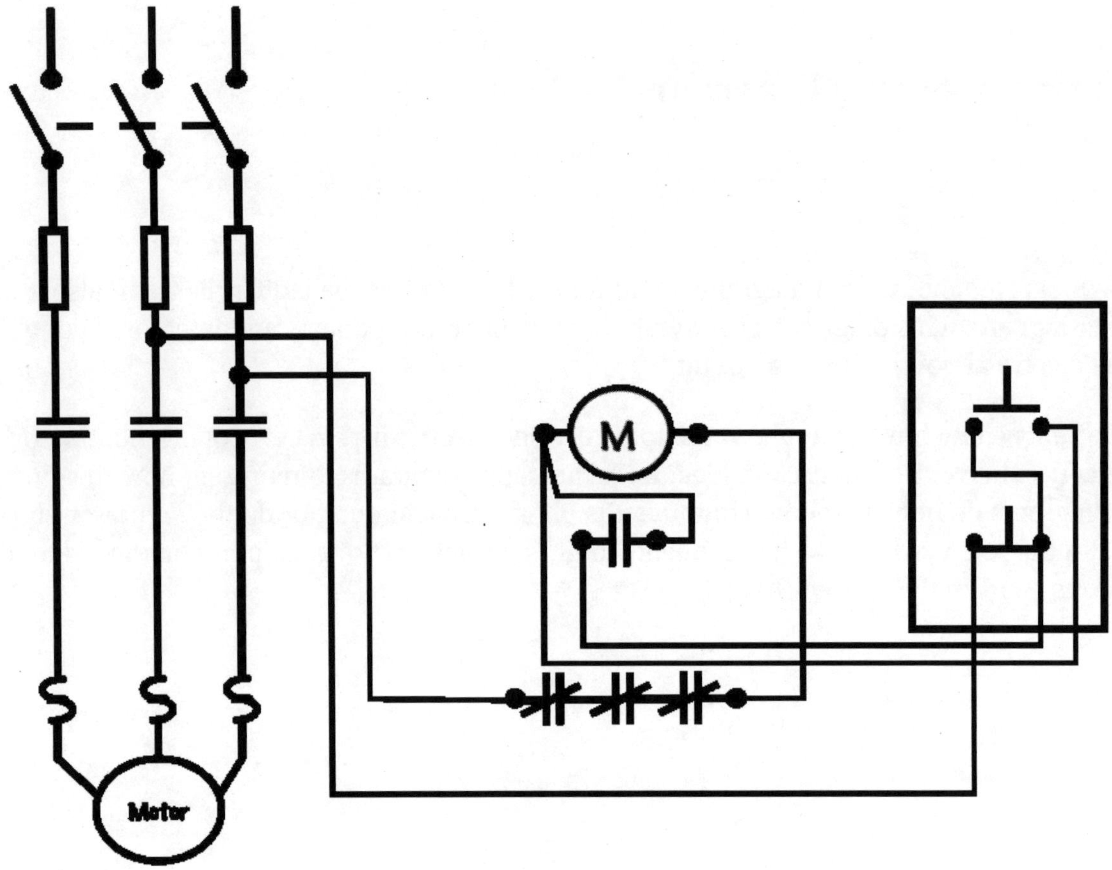

Wiring diagram

20.

Numbering System

Control circuits can range from the very simple to the extremely complex. **Ladder diagrams** show how a system works and are used for troubleshooting purposes. **Wiring diagrams** show where equipment is and how it is connected.

Being able to convert from one style of diagram to the other is an essential skill for electricians and people working with control circuits. To accomplish this, we use a simple **numbering system.**

The numbering system is a way of identifying and "naming" each **electrically common** point in a circuit. A wire is considered "electrically common" with another wire if they share a direct electrical connection with no **switches** or loads between them.

Consider the circuit below:

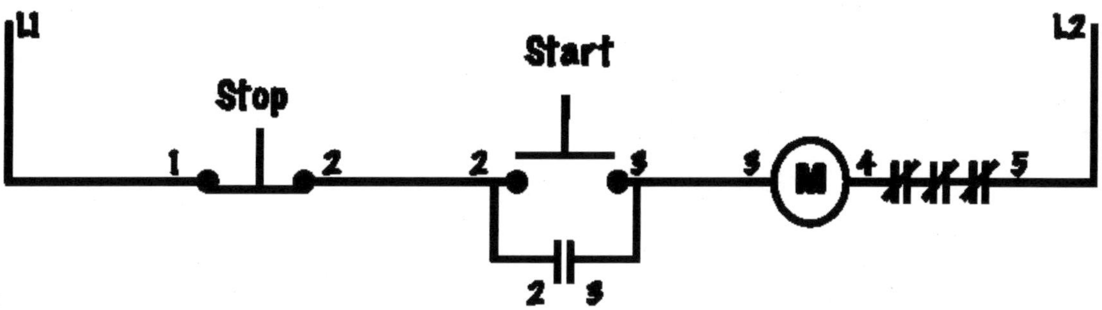

A three-wire numbered circuit

It is a simple three-wire circuit with single stop and start **pushbuttons**. Single-phase power is supplied from Lines 1 and 2 and a motor coil is the load.

Starting at Line 1, we label the first wire "1." When we jump through the first device, (the **normally closed** stop button) we jump up to the next number. Wires 1 and 2 are not electrically common. Coming out of the stop button with wire number 2, we are going to two different places. The first is the **normally open** start pushbutton, the other is the normally open instantaneous contact. Since there are no loads or devices between the load side of the stop button, the line side of the start button, and the holding contact, these points are considered electrically common and can all share the same number. Jumping through these devices, wire number 3 is connected to the load side of both normally open devices and the **motor starter** itself, and so these three points are also electrically common.

Once the control circuit has been numbered, we can use that information to help complete wiring diagrams or determine the number of conductors needed in conduits for installation purposes.

In summary:

- L1 and L2 (or L1 and N) are used to designate the control circuit power.
- Assign number 1 to the first wire entering the control circuit.
- Trace through the **schematic diagram** increasing the numbers given to wires each time you pass through a device.
- Give the same number to all wires that are electrically common, spliced, or connected to the same terminal.

Numbering system
(Schematic to wiring)

A BCcampus element has been excluded from this version of the text. You can view it online here: https://pressbooks.bccampus.ca/basicmotorcontrol/?p=56 (https://pressbooks.bccampus.ca/basicmotorcontrol/?p=56#pb-interactive-content)

21.

Conduit Fill

The **numbering system** can also be used to determine how many wires will need to be pulled through the conduit to connect each device properly.

Consider the following **control circuit** and proposed conduit layout:

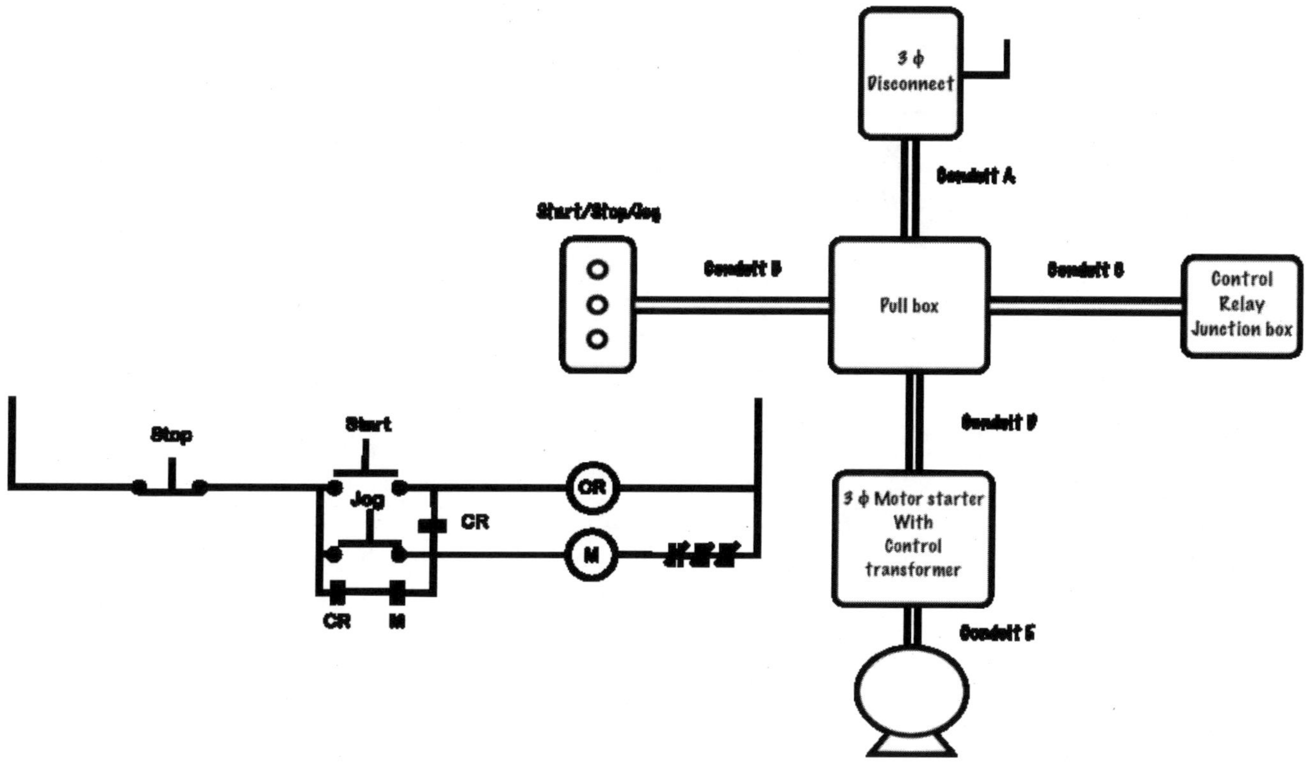

A schematic diagram of a jog circuit utilizing a control relay and the proposed layout of the installation

If this is the proposed physical layout of the equipment, how many wires must be pulled through each piece of conduit?

Conduits "A" and "E" will have only the 3-phase power conductors in them. Conduit "D" will have the three power conductors as well as several control conductors. It is permissible for control conductors to share a conduit with power conductors provided their insulations share the same **voltage** and temperature ratings.

To determine how many control circuit conductors will need to be pulled through each conduit, we apply the numbering system.

First, number the wires in the **schematic**. Each point that is **electrically common** gets the same number. Jump up to the next number each time you go through a device or load.

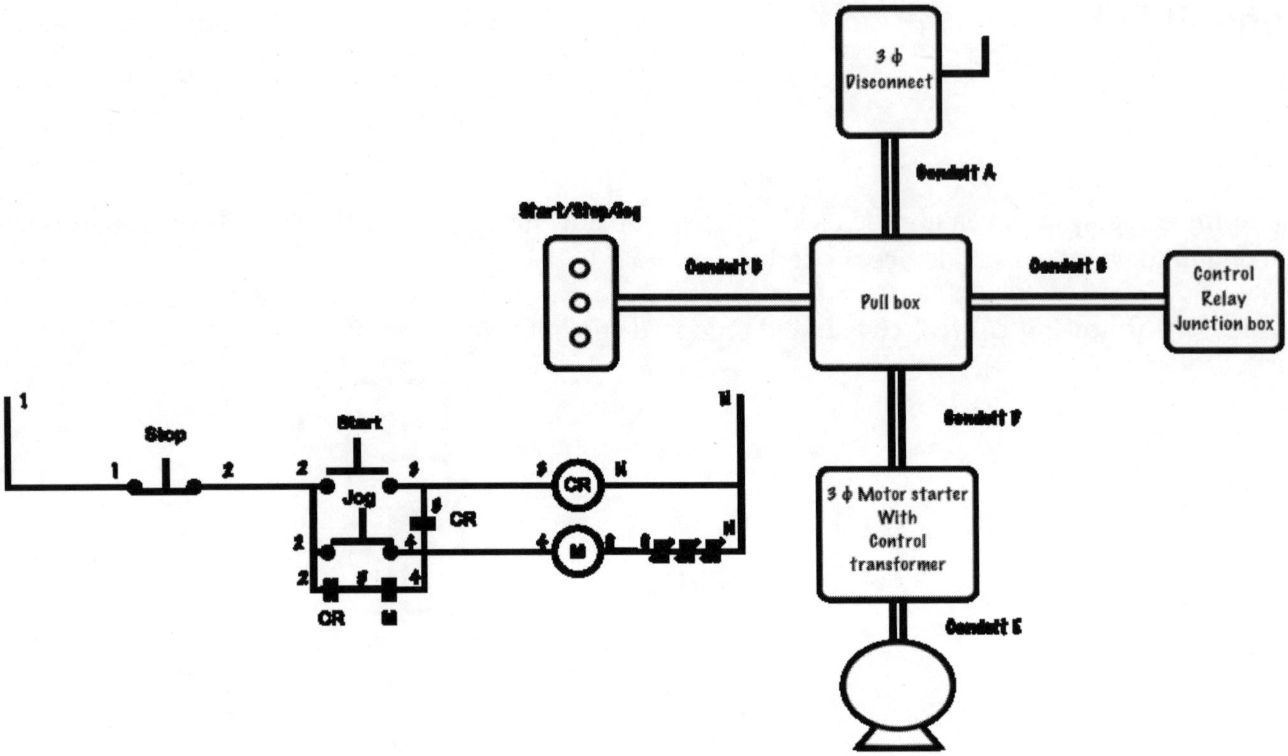

A numbered schematic diagram

Once the schematic diagram is complete, transfer the numbers of each device to their corresponding location in the **wiring diagram**. Each point that has the same number needs a wire connecting them.

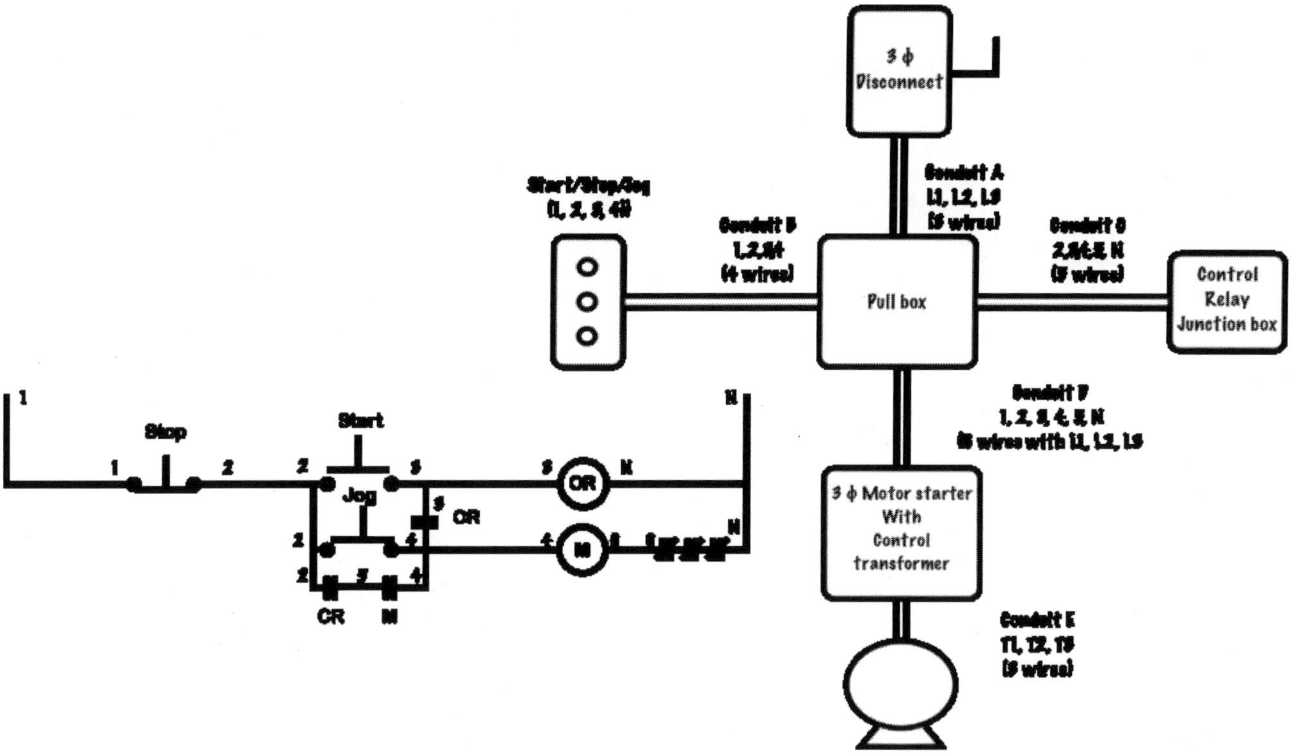

Conduit fill determined in the wiring diagram with numbers transferred from the schematic diagram

Knowing the total number of wires to pull through a complicated run of the conduit is critical because adding wires afterwards can be impractical and difficult.

22.

Transferring From Schematic to Wiring Diagram for Connection Purposes

A completed **wiring diagram** can help with the physical installation of wires. To aid in the development of a wiring diagram, it is useful to start with the circuit's **schematic diagram** and the **numbering system**.

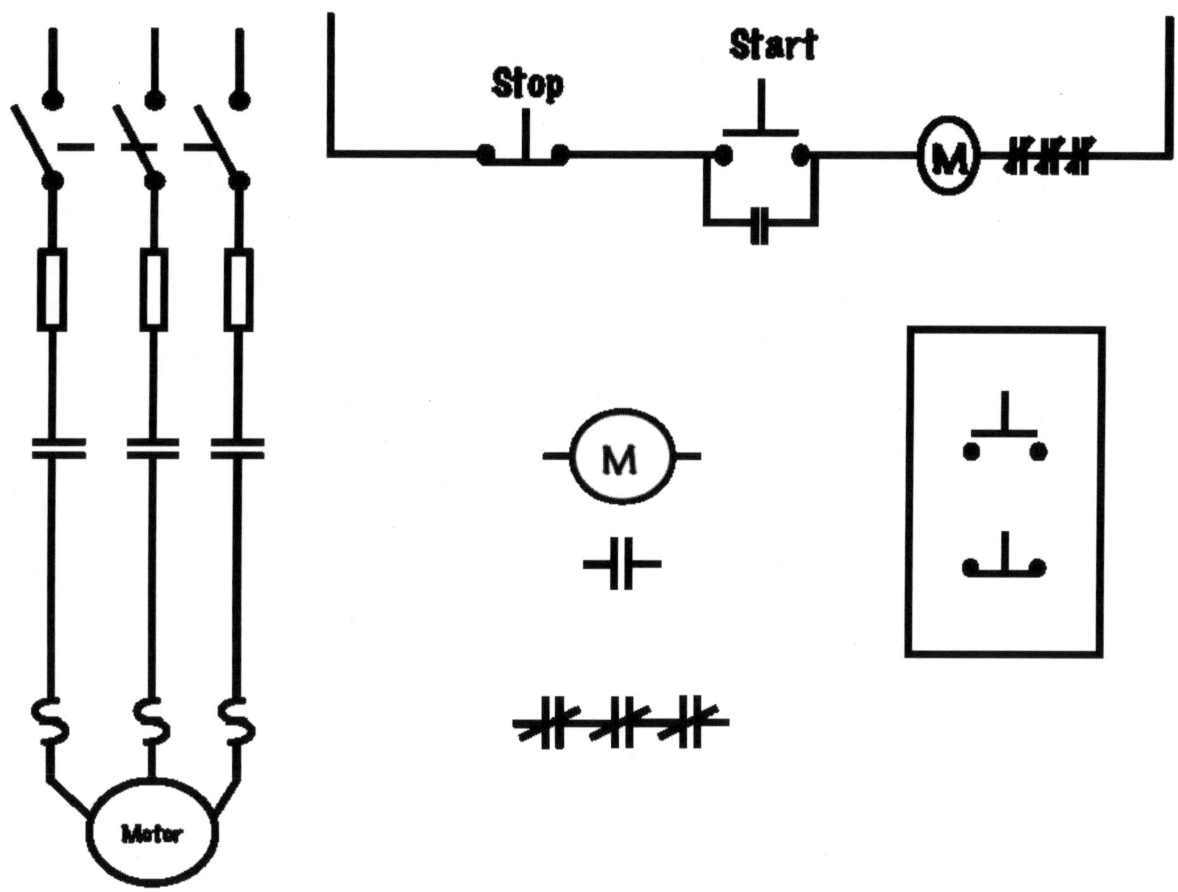

A wiring and schematic drawing

Consider the figure above. It includes a three-wire schematic diagram as well as the equivalent control components and the **power circuit**. In this example, there is no control transformer, so we will be taking control power directly from the line. The **control circuit** power is taken from the load side of the **overcurrent** devices and the line side of the power contacts.

Once the schematic diagram has been properly numbered, each device will have two numbers identifying it, one wire on the line side, and one on the load side. For example, in a three-wire circuit, the stop button gets wires 1 and 2, while the start button and the holding contact each get wire numbers 2 and 3 (hence the term “**2-3 contact**“).

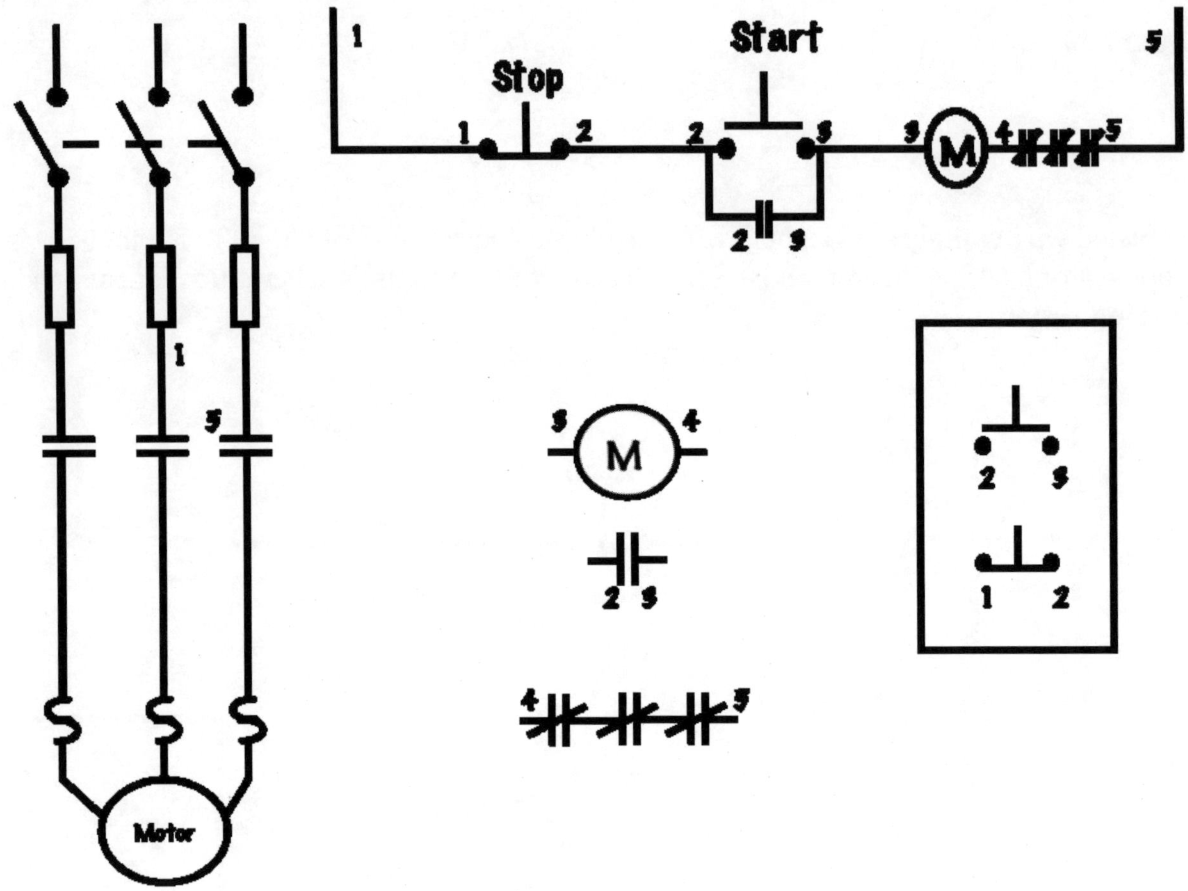

Schematic and wiring diagrams with numbers

Once all the devices have been properly numbered, we simply play connect the dots. Each point that shares the same number is **electrically common** and needs to be connected together. Use straight lines and only connect wires at terminal points on equipment.

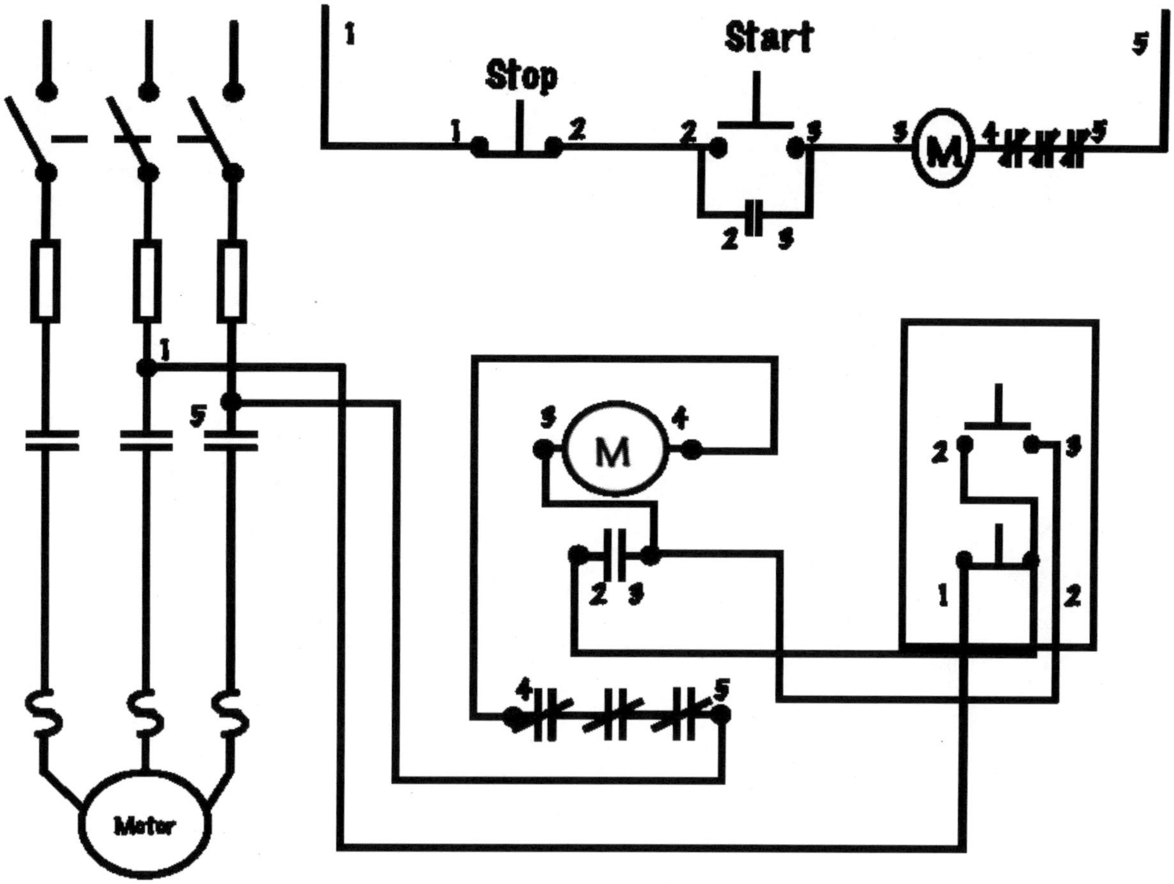

Schematic and wiring diagrams – complete.

Make sure that all connections happen at termination points or "terminal to terminal." In practice, we usually only connect a maximum of two wires to any one point and never make a "free-air splice."

The figure above illustrates each of the strengths that wiring and schematic diagrams have: Schematic diagrams are easy to read and used to logically troubleshoot a circuit, while wiring diagrams show how equipment is physically connected together.

Schematic and Wiring diagrams

A BCcampus element has been excluded from this version of the text. You can view it online here: https://pressbooks.bccampus.ca/basicmotorcontrol/?p=60 (https://pressbooks.bccampus.ca/basicmotorcontrol/?p=60#pb-interactive-content)

23.

Motor-Lead Connections

Three-phase motors use coils of wire to create magnetic fields and produce rotation.

Standard 3-phase motors use six individual coils, two for each phase. The internal construction and connection of these coils inside of the motor is predetermined when the motor is manufactured. There are two classes of 3-phase motors: Wye and Delta.

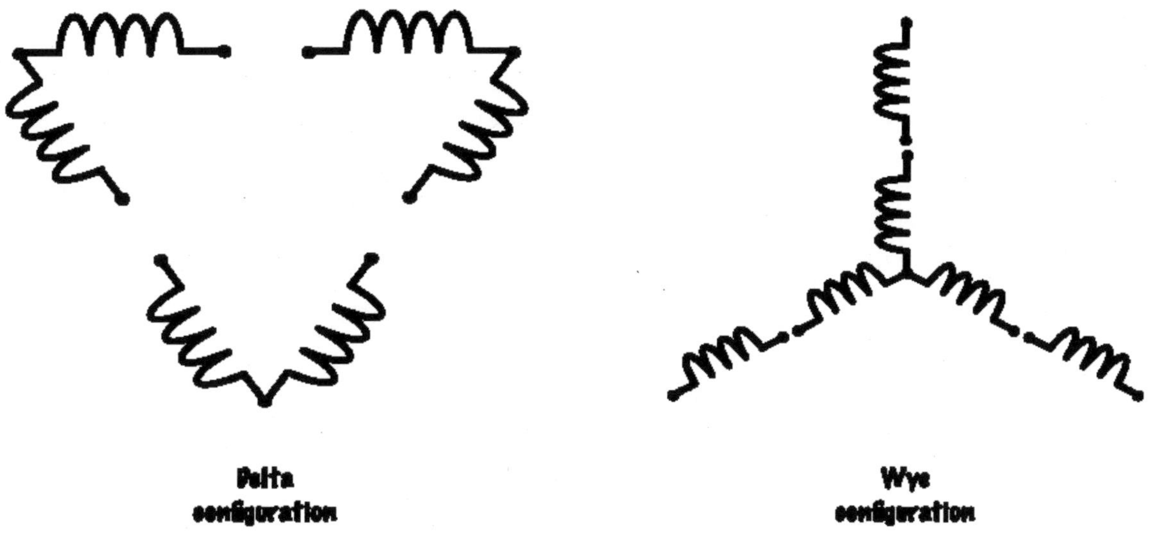

Wye and Delta configuration

Three-phase motors are also constructed to operate at two different **voltages**, and so the coils can be connected in either their high-voltage or low-voltage configurations.

In the high-voltage configuration, the two coils of each phase are connected in **series** with each other so that the higher value of supply voltage is split equally between them and rated current is drawn through each winding.

In the low-voltage configuration, the two coils of each phase are connected in **parallel** with each other so that the lower value of supply voltage is shared equally between the coils and rated current is drawn through each winding.

Note that the low-voltage connection will necessarily need to draw twice as much current from the source as the high-voltage connection. Most motors will list two values of voltage and current on their nameplates. It is important to size **motor starters** and their **overload relays** based on the expected value of current that is to be drawn by the motor at the voltage it is used at.

Each of the six individual coils has two leads supplying it, for a total of twelve leads in total. In both the Wye and Delta configurations, three of these leads are connected internally, and so only nine leads are brought out of the motor for connection. These leads are numbered 1–9, and in both Wye and Delta follow a standard numbering convention: starting at the top of the diagram with wire number 1, draw an inwardly descending spiral from each connection point, ascending to the next number at each step.

Depending upon the internal construction of the motor, these leads can be hooked up in one of four ways: High- or low-voltage Wye, or high- or low-voltage Delta

Identifying Wye or Delta with an Ohmmeter

It sometimes becomes necessary to test or confirm the configuration of a motor before final connection. If a Wye wound motor is connected as a Delta wound motor or vice versa, the motor will not operate properly.

Consider this situation: You have nine leads coming from a motor, but no indication of whether its Wye or Delta wound. By using an **ohmmeter** to do a simple continuity check, you can determine the construction type of the motor.

If it is Wye wound, each of wires 1, 2, and 3 should only have continuity with one other lead (4, 5 and 6 respectively). The three leads without continuity to wires 1, 2, and 3 should all have continuity with each other.

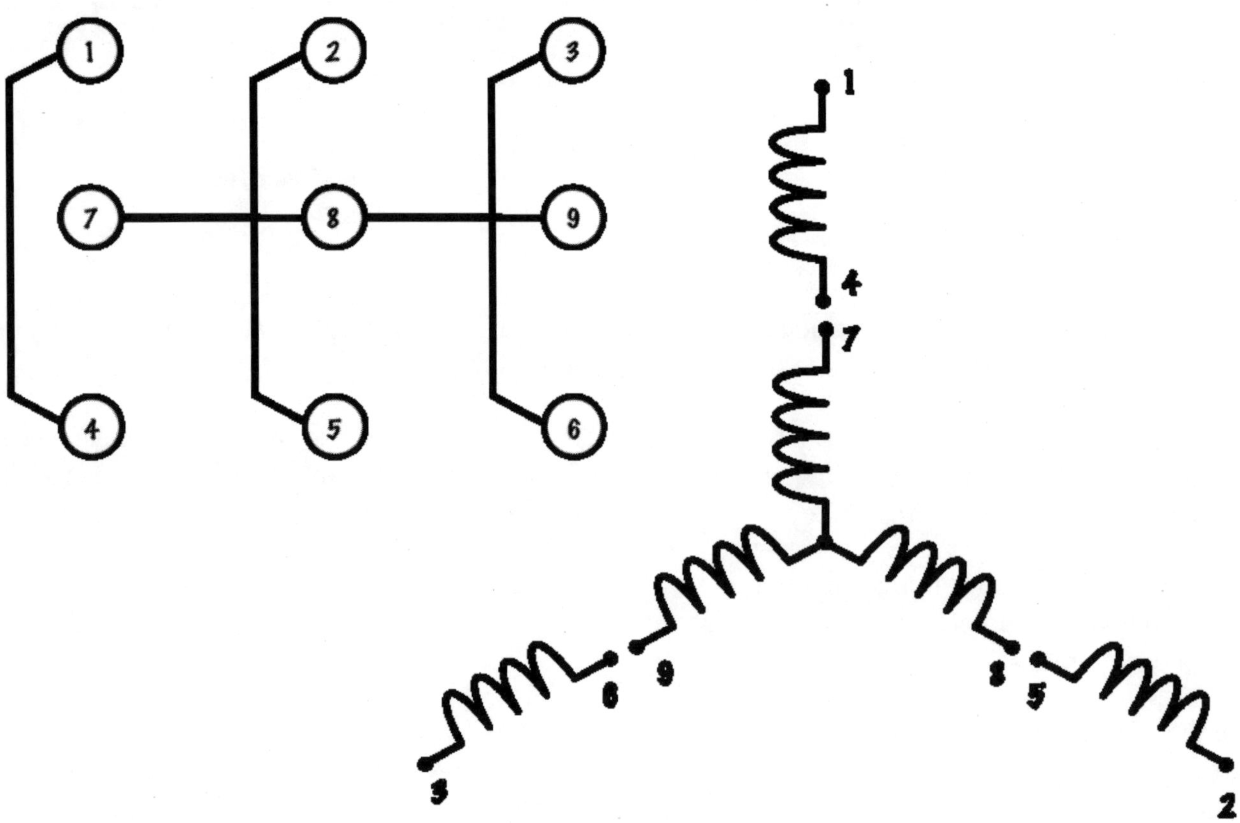

Wye motor connections

If it is Delta wound, each of wires 1, 2, and 3 should have continuity with two other leads:

- T1 has continuity to T4 &T9
- T2 has continuity to T5 &T7
- T3 has continuity to T6 &T8

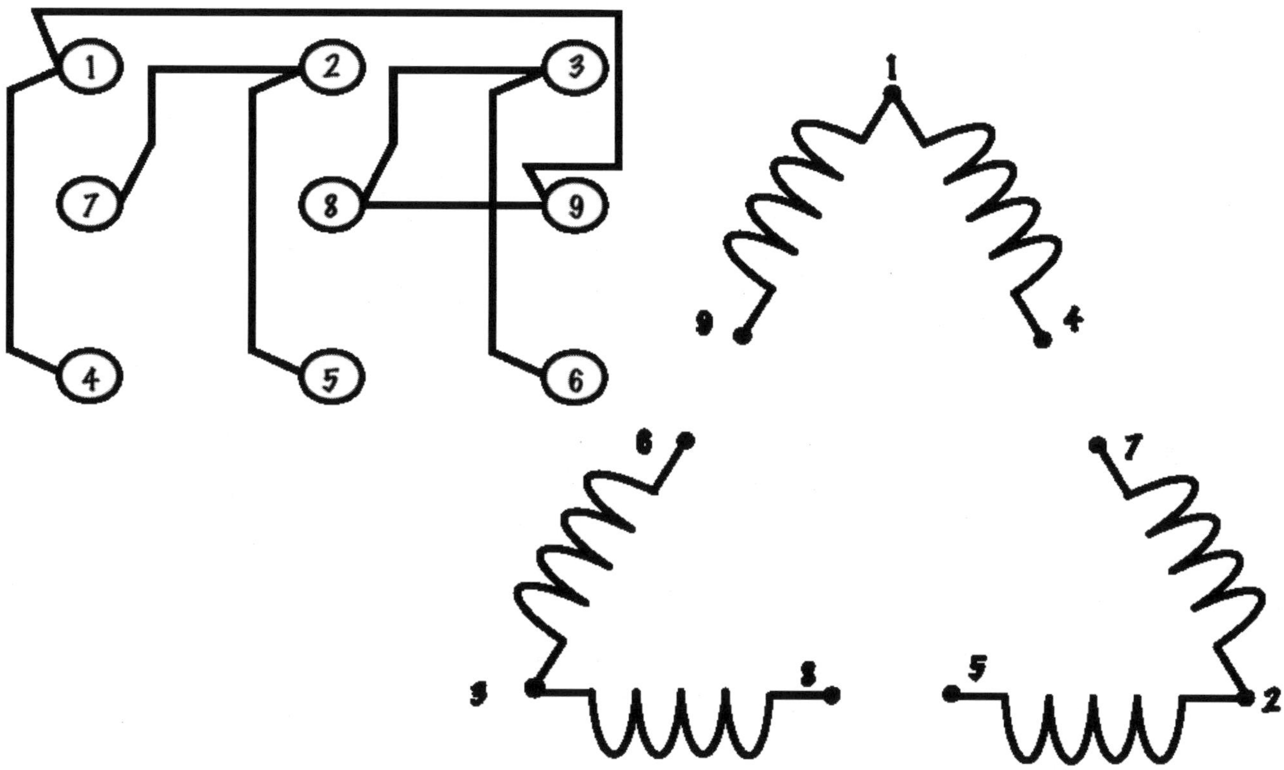

Delta Motor Connections

It is important to note that these points represent the internal connection of the coils of the motor, not how they are to be hooked up to voltage.

Wye Connections

Low-voltage Wye

In this configuration, each phase is brought to two coils that are connected in parallel with each other. Terminals 4, 5, and 6 are tied together to make a second neutral connection.

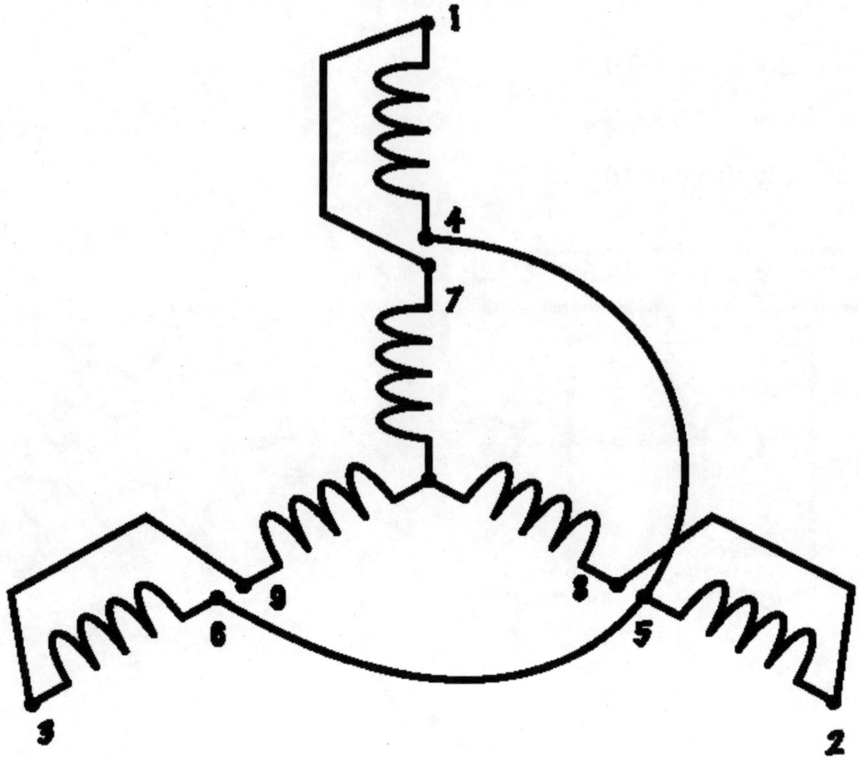

Low-voltage Wye connection

L1	L2	L3	Tie together
1,7	2,8	3,9	4,5,6

High-voltage Wye

In this configuration, each phase is brought to two coils that are connected in series with one another.

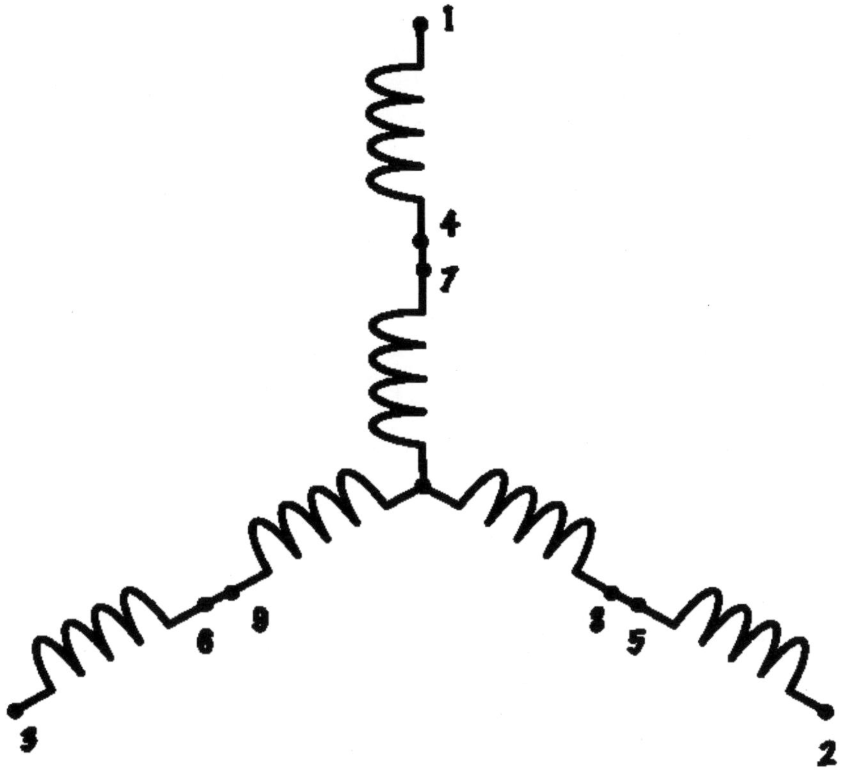

Wye motor high-voltage connection.

L1	L2	L3	Tie together
1	2	3	4,7 – 5,8 – 6,9

Delta Connections

Low-voltage Delta

In this configuration, each phase is brought to centre connection of two coils and to the end connections of each of the other two groups of coils.

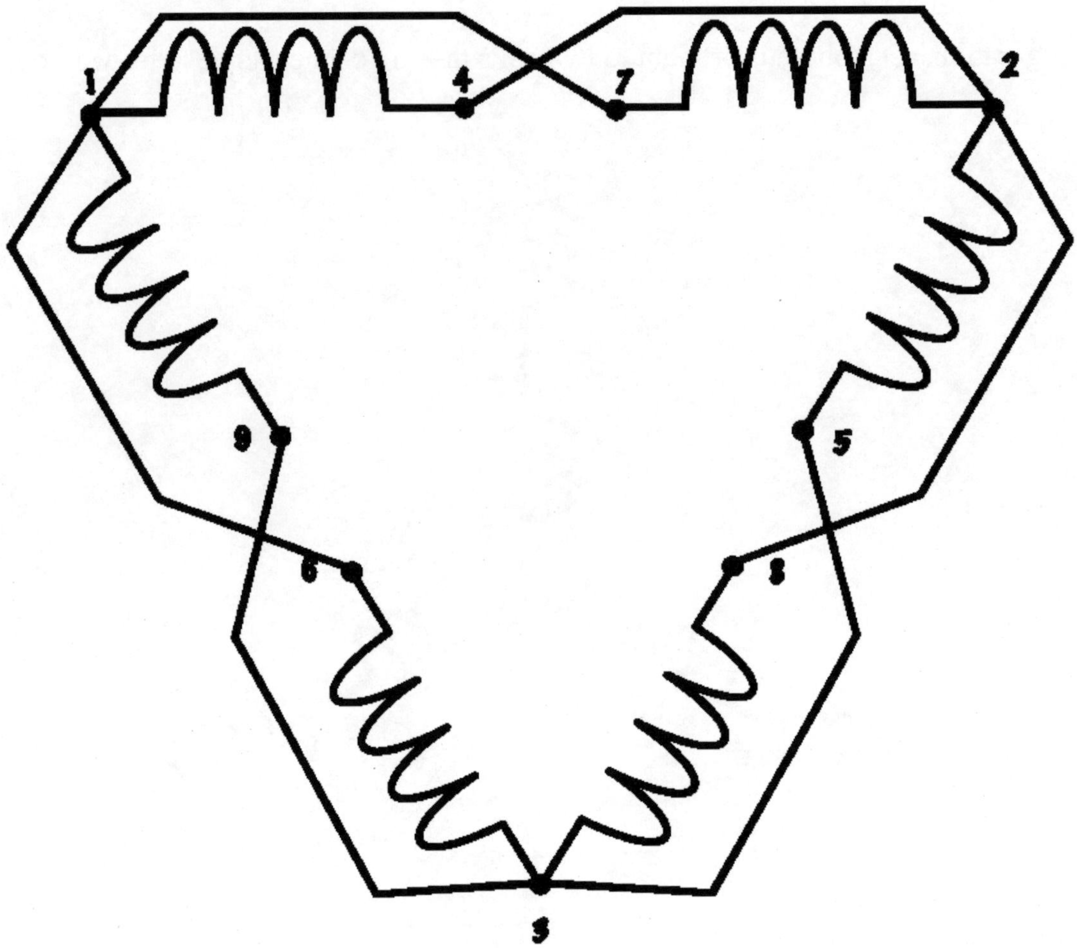

Delta Motor Low Voltage Connection

L1	L2	L3	Tie together
1,6,7	2,4,8	3,5,9	none

High Voltage Delta

In this configuration each phase is brought to two coils that are connected in series with the other phases coils.

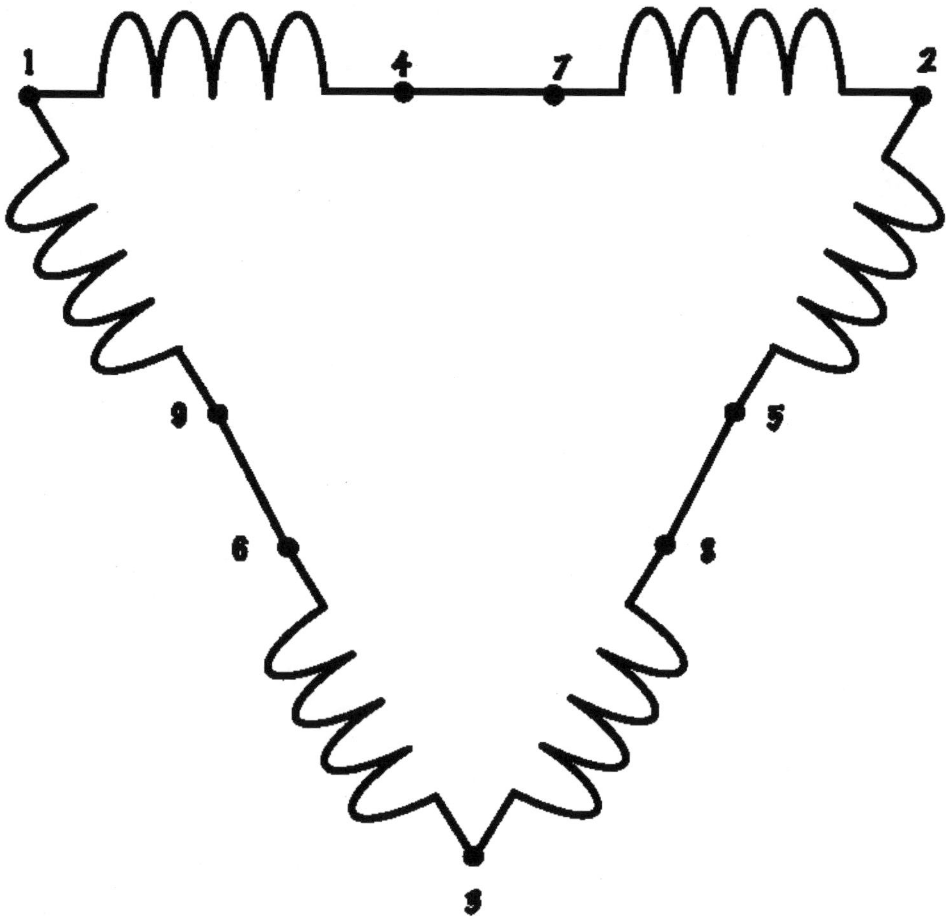

Delta Motor High Voltage Connection

L1	L2	L3	Tie together
1	2,	3	4,7 – 5,8 – 6,9

24.

Self Test 4

An interactive or media element has been excluded from this version of the text. You can view it online here: https://pressbooks.bccampus.ca/basicmotorcontrol/?p=519 (https://pressbooks.bccampus.ca/basicmotorcontrol/?p=519#pb-interactive-content)

If using the print, PDF, or eBook copy of this book, navigate to the above link to complete the quiz. However, the quiz questions are also provided at the end of the book for offline use: Offline Copies of Chapter Quizzes

V

Circuits

Learning Objectives

After studying this section, students should be able to:

- Identify and differentiate between a power and a control circuit.
- Describe various sources of supply for control circuits.
- Describe and identify common two-wire automatic circuits.
- Describe and identify common three-wire manual circuits.
- Describe various jog circuits.
- Describe various forward/reverse circuits and their safety components.
- Describe various circuits which utilize timer relays for sequence control.

25.

Circuits Overview

When discussing motor controls, there are two circuits that need to be well understood.

The first is the **power circuit**. This circuit is relatively straightforward. It is a 3-phase circuit which powers the motor and can operate as high as 600A at 600V or even beyond.

The second is the **control circuit**. This is the circuit which controls the power circuit. This circuit can be as simple as a two-wire **pilot device**, or it can be very complex with many interlocking features and multiple starter and timer coils. The control circuit will often operate at a lower **voltage** relative to the power circuit.

We will be spending the majority of our time discussing common control-circuit designs, so it's best to start with a quick overview of the power circuit.

26.

Power Circuit

The **power circuit** of a three-phase motor consists of the following items:

- 3PST three-phase disconnect
- **Overcurrent** protection
- **Horsepower-rated motor starter contacts**
- Overload relay heaters

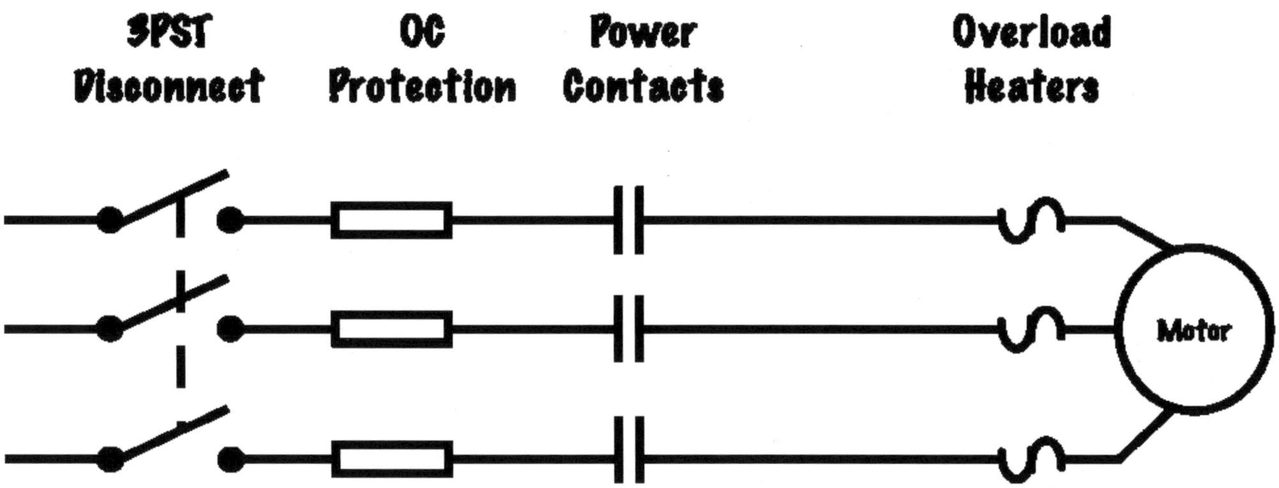

In practical applications, these four components would often be mounted in a combination starter, which might house a **circuit breaker** and **motor starter** in the same cabinet.

In order for the motor to operate, the 3-phase disconnect must be closed, the **fuses** must be in good working order and the power contacts must be closed. The start and stop action of the motor is determined by the **control circuit**, which energizes the magnetic motor starter, pulling in or dropping out the **armature** that then opens or closes the power contacts.

The current that goes to the motor then runs through the heater elements of the **overload relay**. If the motor draws too high a current for too long, the heater element causes a **normally closed contact** to open, which then interrupts the current to the motor starters coil. This causes the power contacts to automatically open and turn off the motor.

This relationship also provides **low-voltage protection (LVP)** in the case of a power outage. Motors controlled from magnetic starters need to be manually reset following a period of low **voltage**.

The motor starters that we will be focusing on are "across-the-line" magnetic motor starters. This refers to them using full-line voltage to start the motor.

Across-the-line motor starters draw large **inrush currents** when energized. To minimize inrush currents and severe starting torques, some circuits use a variety of "soft-start" devices, but these devices are beyond the scope of this text.

Power circuits are identical for most 3-phase motor starters, so it is common not to refer to them in **schematic drawings**.

27.

Control Circuit's Source of Supply

Compared to the **power circuit**, the **control circuit** is where things get interesting.

Control circuits can be simple two-wire on/off devices that energize a **motor starter**, or they can be more involved **three-wire circuits** with multiple **pushbutton** locations and motors that start in sequence.

One of the key advantages of control circuits is that they can operate at a lower **voltage** than the motor that they control. This allows an operator to push a 120V pushbutton, which then energizes a 120V starter coil whose **contacts** then close on a 600V motor. If a dangerous fault occurs, the high-voltage switching happens in a location away from the operator.

Since the value of voltage between conductors in a 3-phase system could be as high as 600V and the control circuit is usually a single-phase 120V AC circuit, some form of control-circuit voltage must be available. Power for the control circuit can be supplied in one of three different ways.

A Separate Source

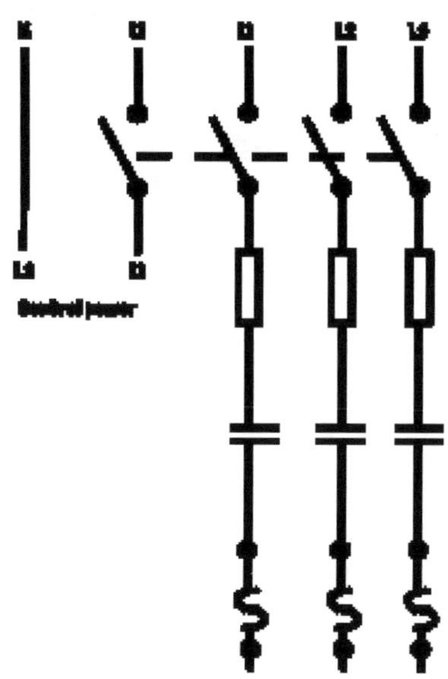

Control Power Separate Source

This system can provide a lower voltage for the control circuit, but requires additional wires and breakers to be installed at the main-load centre to supply the control circuit.

Direct from the Line

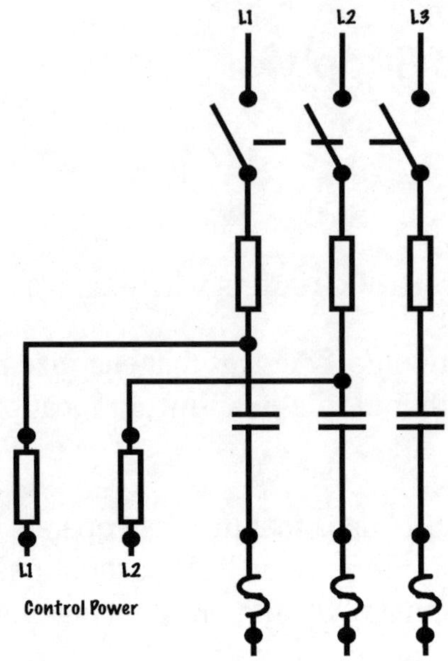

Control Power Direct from the Line

This system does not allow the control circuit to operate at a lower voltage than the power circuit. The advantage is that no additional circuits need be drawn from a load centre. Power is taken directly from the power circuit after the three-phase disconnect and **overcurrent** protection. This ensures that if either of these opens, the control circuit will also be disconnected from its source of supply.

A Control Transformer

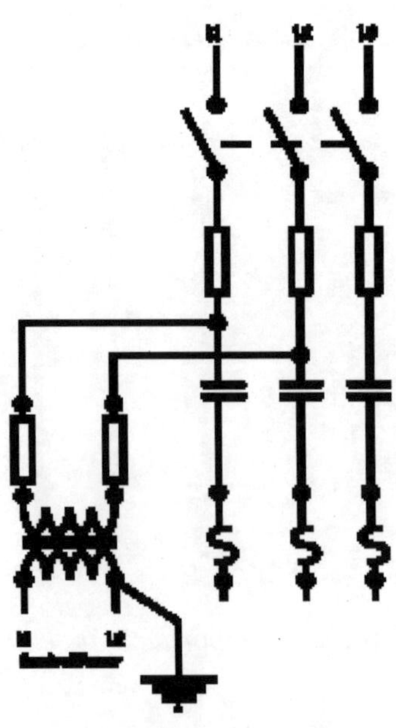

Control power with control transformer.

The connection is similar to when power is taken directly from the line, but instead of connecting directly to the control circuit, the line voltage is first stepped down, usually to 120V AC. The secondary of the control transformer then acts as the source of supply for the control circuit and must be of sufficient size to supply power to the motor starter and any other associated contactor equipment such as **control relays**, timers or **pilot lights**.

A BCcampus element has been excluded from this version of the text. You can view it online here: https://pressbooks.bccampus.ca/basicmotorcontrol/?p=68 (https://pressbooks.bccampus.ca/basicmotorcontrol/?p=68#pb-interactive-content)

28.

Automatic Control Circuits

Also known as **two-wire circuits**, automatic **control circuits** can be controlled manually from simple two-wire ON/OFF toggle **switches** or any number of two-wire **pilot devices** such as temperature switches or float switches.

If there is a power failure while the switch is in the ON or closed position, the motor will automatically restart when power returns. This is an example of **low-voltage release (LVR)**. Some examples include sump pump circuits and temperature circuits.

Two-Wire Temperature Circuit

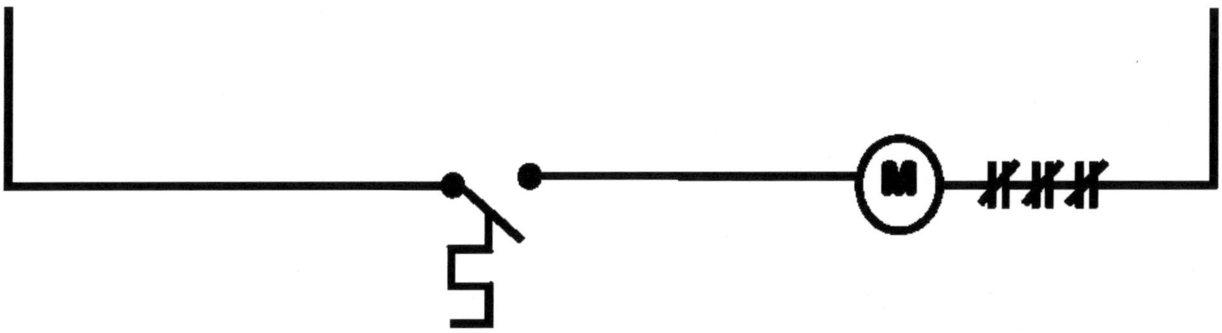

Two-wire control circuit

The diagram below has two components: the normally open temperature switch and the magnetic contactor. The three normally closed contacts that represent the OLR protection are connected in **series** with the motor coil and so can be treated as a single device. In practice, a **motor starter** often only has two external connection points and all internal **overload contacts** connected in series. This arrangement allows for the control of the magnetic contactor by the opening and closing of the temperature switch.

The switch can be housed in a separate enclosure from the one housing the motor contactor, which is typically close to the motor it controls or in a motor-control centre (MCC). Separate housing of the pilot device that controls the load is useful for remotely monitoring situations. When wiring these pilot devices, it is useful to refer to the **schematic diagrams** and the **numbering system**.

This circuit takes two wires: One constant hot from the source and one switch leg to control the load, hence two-wire control.

If the ambient temperature causes the switch contacts to close while there is a power failure, the load will automatically re-energize when power returns, which is an example of low-voltage release (LVR).

In practicality, the temperature switch could be replaced with any pilot device and the essential nature of the circuit wouldn't change. The main limitation of using only a single-pilot device to control the load is the range of sensitivity of that single pilot device.

Consider the next circuit which uses two float switches to control a pump motor.

29.

Sump-Pump Circuit

A three-position selector **switch** may be used to provide either manual or automatic control. When the switch is moved to the HAND position, the magnetic starter is energized. This is useful for testing the condition of the motor and verifying the direction of rotation. When the selector switch is moved to the AUTO position, the magnetic starter is controlled by the opening or closing of the float-switch contacts.

The diagram below shows a sump-pump circuit, while the figure represents the tank that it empties.

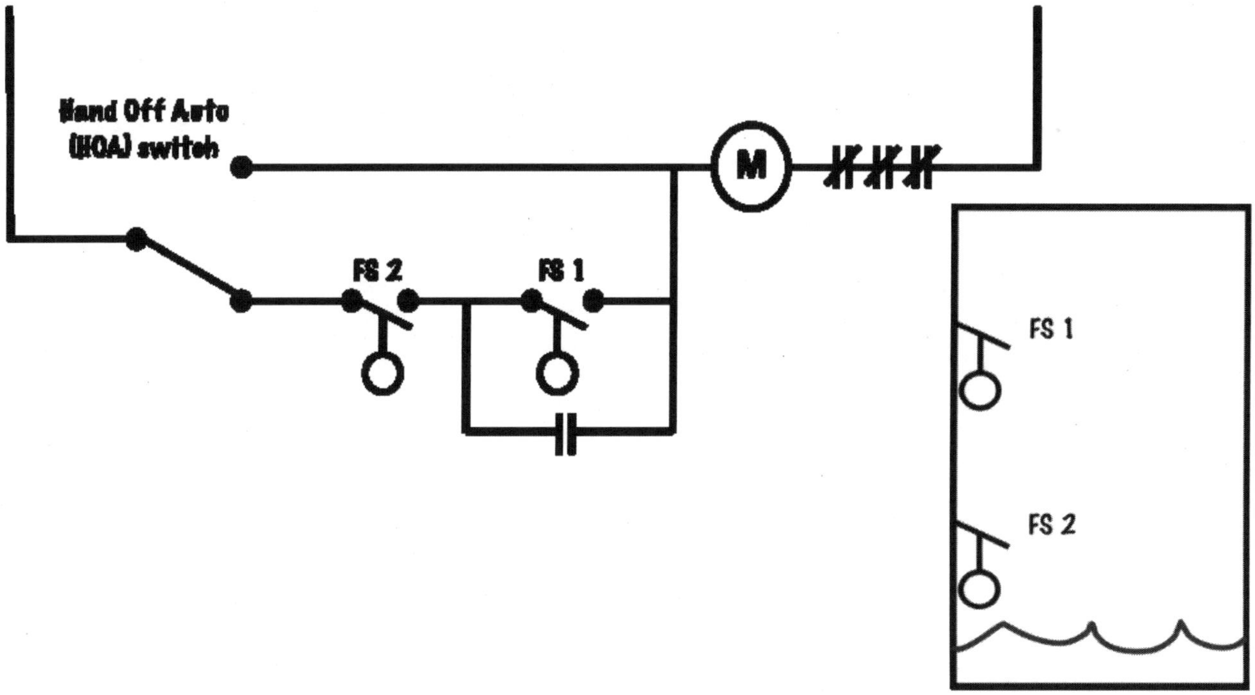

Sump-control circuit

If the water level is low, both float switches are **normally open** and the motor does not run. If the water level rises, FS 2 will close first, but the motor will not start. Current cannot get past FS 1 nor the normally open **auxiliary contacts**. If the water level rises to fill the tank and FS 1 closes, the motor will energize and start pumping water out of the tank and very quickly FS 1 will open again.

This is where having two **pilot devices** allows us a greater range in sensitivity. Even though FS 1 opens almost immediately, the auxiliary contact keeps the motor running until the tank is fully emptied and FS 2 opens. With only a single pilot device the motor would only drain a small portion of the tank and would be subjected to multiple starts and stops. Using two pilot devices to control the motor allows for more efficient operation of the motor.

This circuit shows the arrangement for a tank-drain function. A simple rearrangement of **contacts** can provide a tank-fill function as well.

Sump Control Circuit

A BCcampus element has been excluded from this version of the text. You can view it online here: https://pressbooks.bccampus.ca/basicmotorcontrol/?p=74 (https://pressbooks.bccampus.ca/basicmotorcontrol/?p=74#pb-interactive-content)

30.

Reservoir Circuit

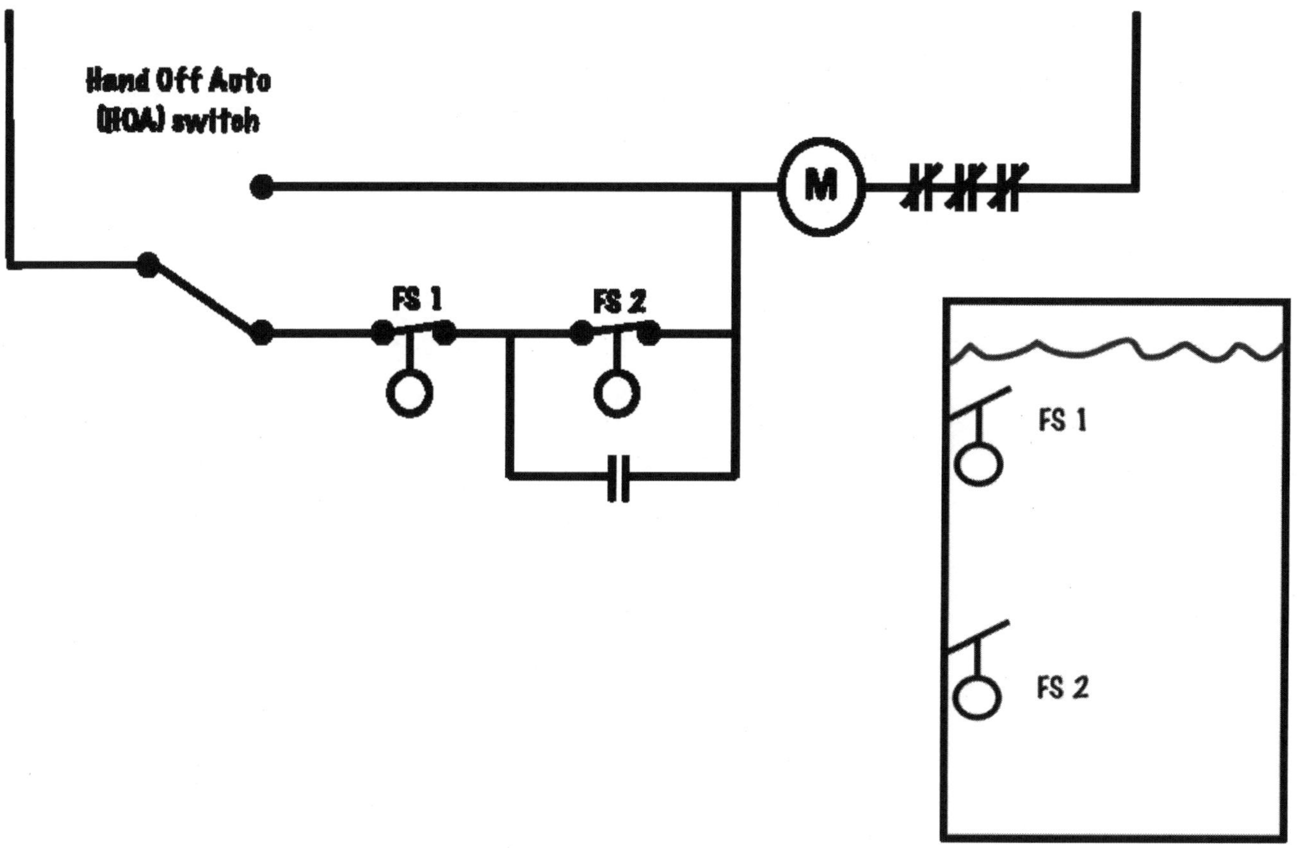

Reservoir Control Circuit

A three-position selector **switch** is again used to provide either manual or automatic control. When the selector switch is moved to the AUTO position, the magnetic starter is controlled by the opening or closing of the float switch contacts, similarly to the sump-pump circuit previously discussed.

The key change is in the nature of the switches contacts. For a tank-fill circuit, our float switches will have their contacts wired normally closed. This means that both switches will close at a low liquid level and open at a high level. Also note that **auxiliary motor contact** is now in **parallel** with FS 2 or the low-level switch. This ensures that the motor starts pumping liquid into the tank when it's nearly empty and doesn't stop until the high-level switch is opened.

Again, with two **pilot devices**, our circuit has a greater range of sensitivity than it would with only a single float switch.

Reservoir control circuit

A BCcampus element has been excluded from this version of the text. You can view it online here: https://pressbooks.bccampus.ca/basicmotorcontrol/?p=72 (https://pressbooks.bccampus.ca/basicmotorcontrol/?p=72#pb-interactive-content)

31.

Manual Circuits

A manual circuit is one that is stopped and started by a person. These circuits might have automatic features (such as overload protection) but their main feature is that they require some sort of input from a human operator.

The most fundamental manual circuit is the stop-start **three-wire circuit**. A firm understanding of this circuit is necessary because many of the features of the three-wire circuit scale upwards to more complex circuits.

Basic Three-Wire Circuit

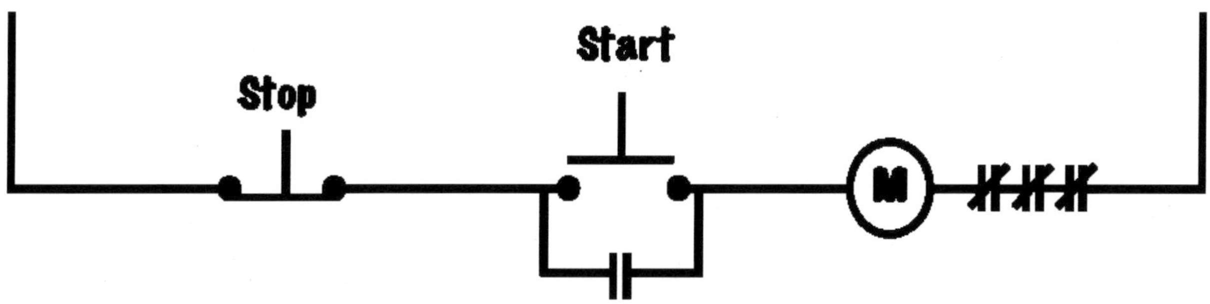

A basic three-wire schematic drawing

The components in the circuit above consist of a **normally closed**, momentary-contact stop **pushbutton**; a **normally open**, momentary-contact start pushbutton; and a **motor starter**. These components may be housed in the same compartment, or the pushbuttons could be installed separately from the motor starter. The motor starter itself is connected in **series** with the three normally closed **overload relay** contacts.

The unique component of the circuit, indeed the device that allows the circuit to provide **low-voltage protection (LVP)**, is the normally open instantaneous contact connected in **parallel** with the start button. The normally closed stop button offers no opposition to the flow of current.

When power is on and the motor is not running, current can pass neither the normally open start button, nor the normally open contact, and so the motor does not start.

When an operator comes and pushes the start button, current is able to pass through the **contacts** as the button is depressed. This completes the circuit and energizes the coil of the motor starter, pulling in the **armature**. The three **horsepower-rated contacts** will close, as well as the normally open **auxiliary contact** in parallel with the start button.

The normally closed contacts in series with the coil are controlled by the thermal action of the overload relays, and not the movement of the armature, so they remain closed as long as an **overload** does not occur.

Once the normally open contact (called a holding, maintaining or 2 – 3 contact) is closed, the operator is now free to release the start button. The current will continue to have a path through the circuit, keeping the starter coil energized. In reality this all happens in a fraction of a second, and so the holding contact will be closed before the operator's finger has come off the start button.

Now the motor will continue to run until current to the motor starter is interrupted. This could happen if the operator presses the normally closed stop button or if an overload occurs and one of the normally closed OLR contacts opens up. If a power outage occurs during normal operation, the starter coil will drop out and the motor will stop as well.

In any situation where the motor starter becomes de-energized and the armature is pushed open by the spring inside it, all contacts associated with the coil return to their normal condition. The motor will not restart without further input from an operator as the holding contact will have returned to its normally open state.

With the installation of an auxiliary contact in parallel with the momentary start button, any circuit can provide low-voltage protection (LVP).

Start/Stop Pushbutton
circuit
(Basic 3 wire)

A BCcampus element has been excluded from this version of the text. You can view it online here: https://pressbooks.bccampus.ca/basicmotorcontrol/?p=82 (https://pressbooks.bccampus.ca/basicmotorcontrol/?p=82#pb-interactive-content)

32.

Three-Wire Circuit with Multiple Push Buttons

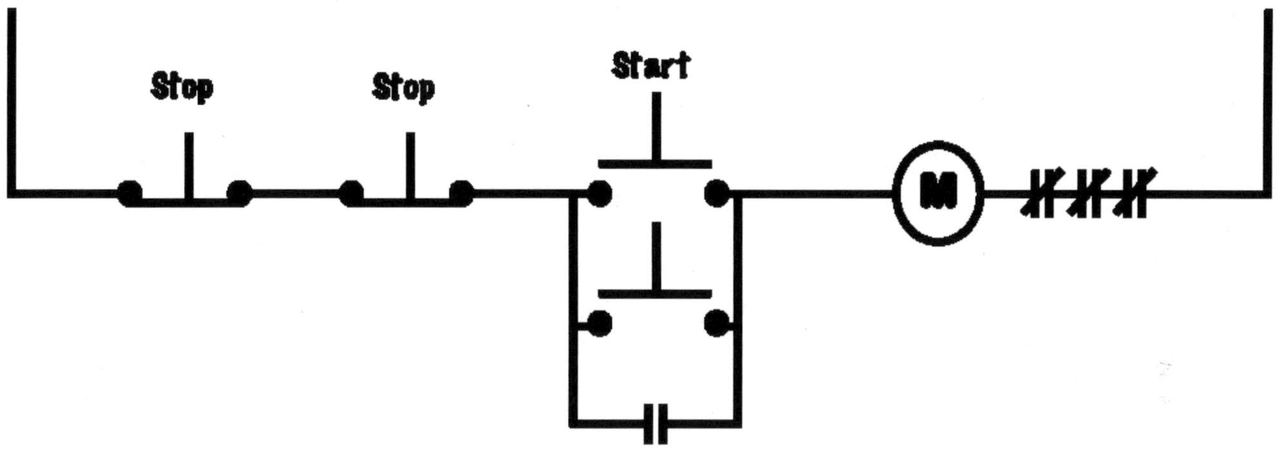

Multiple pushbutton circuit

This circuit is functionally identical to the **three-wire circuit** previously discussed, but allows control of the motor from two different locations, often called "local" and "remote" **pushbutton** stations.

As a rule, a motor can be controlled from any number of locations provided that we follow these simple rules:

1. Wire all **normally closed** stop buttons in **series** with each other,
2. Wire all **normally open** start buttons in **parallel** with each other, and
3. Connect the holding contact in parallel with the start buttons.

For the above example, we will only consider two push-button stations, but the principle can be scaled up to include any number of pushbutton stations.

With all the control current going through the normally closed contacts of the stop buttons, pushing any one will interrupt the circuit and disconnect the **motor starter**. Pushing any start pushbutton will provide a path for current to energize the motor starter.

Once energized, the **2-3 holding contact** will close, allowing the start button to be released, and the circuit will now run while providing **low-voltage protection (LVP)** until either an **overload** occurs or a stop button is pressed.

Multiple pushbutton stations in a 3wire circuit

A BCcampus element has been excluded from this version of the text. You can view it online here: https://pressbooks.bccampus.ca/basicmotorcontrol/?p=84 (https://pressbooks.bccampus.ca/basicmotorcontrol/?p=84#pb-interactive-content)

33.

Three-Wire Circuit With Pilot Lights

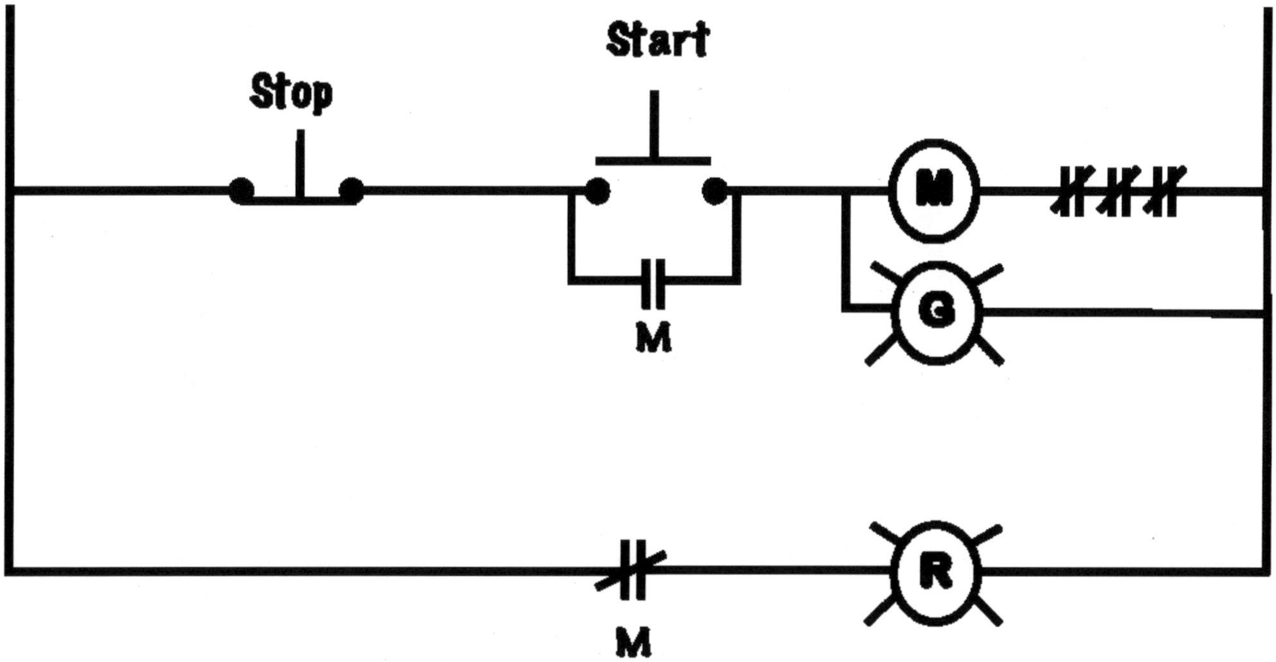

Three-wire control circuit with pilot lights

Motors are often installed in loud environments or at remote locations, and so simply relying on the sound of a running motor to tell an operator its status is unreliable. To aid with this, we can connect **pilot lights** in our **control circuit** to indicate the running status of a motor. Pilot lights that are installed in the control circuit in **parallel** with the motor starter will be energized when the motor is energized. These are often referred to as motor running lights.

Pilot lights must NEVER be installed in **series** with motor starters; the motor starter might not have enough **voltage** to pull in, and if the filament in the pilot light burns out, it will create an open in series with the rest of the circuit.

Motor starters often come with **auxiliary contacts** beyond the **normally open 2-3 holding contacts**, and if not, many starters can have additional auxiliary contacts added to them.

If a motor starter has a set of **normally closed contacts** these can be wired in series with a pilot light. As long as the motor starter is not energized, the contacts will remain closed and the light will be on. This can indicate to an operator that a motor is in the off or de-energized condition.

34.

Pressure Circuit

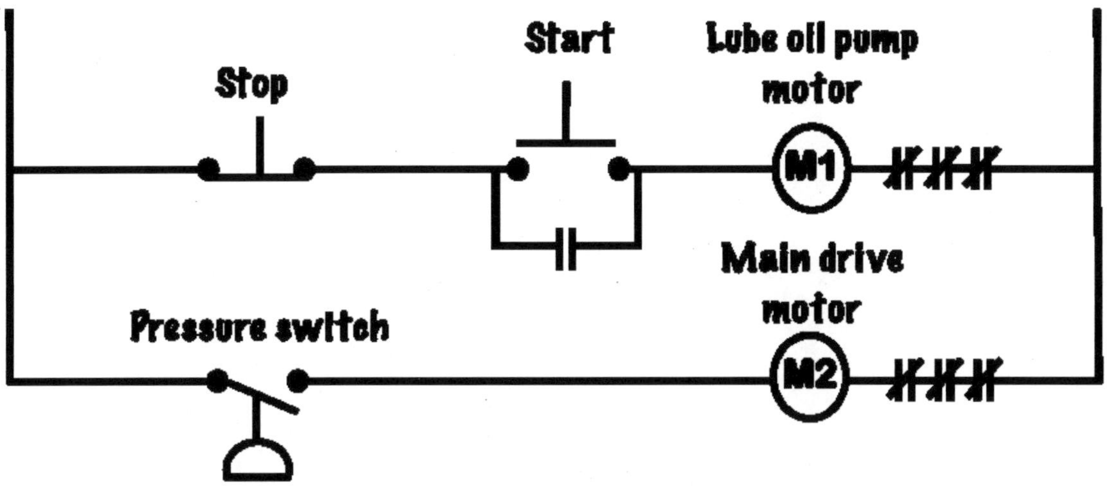

Lube oil pump system

This looks like a standard **three-wire circuit** for driving two motors, but upon closer examination, we see that **motor starter** M1 is controlling a motor pump for lubricating oil. The main drive motor, controlled by motor starter M2 is controlled by a **normally open** pressure switch.

By pressing the start button, motor M1 will start to pump lube onto the bearings of the main motor. Once the pressure meets the minimum threshold and closes the pressure switch contacts, motor M2 will be allowed to start. This arrangement ensures the motor M2 will not run if there is insufficient lubrication pressure.

The circuit will continue to operate as a standard three-wire circuit providing **low-voltage protection (LVP)** until either the stop button is pressed, or an **overload** occurs. If an overload occurs in motor M1, then M2 will continue to run until the pressure switch drops open. If an overload occurs in M2, it will stop while motor M1 will continue to keep lubrication pressure up.

Lube oil pump
System

A BCcampus element has been excluded from this version of the text. You can view it online here: https://pressbooks.bccampus.ca/basicmotorcontrol/?p=76 (https://pressbooks.bccampus.ca/basicmotorcontrol/?p=76#pb-interactive-content)

35.

Jog Circuits

Sometimes called "**inching**," jogging is the term given to the momentary energization of a motor only so long as an operator is pressing a button.

A **jog circuit** is a circuit that allows an operator to either start the motor or "**jog**" the motor and are commonly used for motors controlling conveyor belts to allow for precise positioning of materials.

Any **motor starter** that is used to jog a motor will be subjected to repetitive **inrush currents**, which can cause overheating of the power contacts. If a motor is expected to be jogged more than five times in a minute, the motor starter should be increased in size and horsepower rating for this more severe operating condition.

To achieve a jog function, there are several common circuit designs, each with their own advantages and disadvantages. A common feature that all jog circuits have is that they have some method of disabling the **holding contact** used in the **three-wire circuit**. This is usually accomplished by putting some component in **series** with the holding contact, such as a **switch** or momentary **pushbutton**.

Jog Circuit with Selector Switch

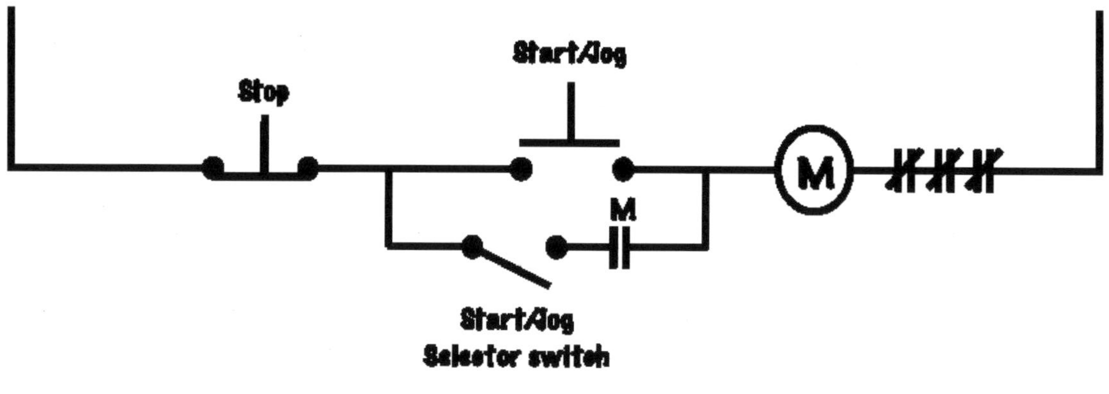

Selector switch jog control circuit

The most basic of the jog circuits, this is essentially a three-wire circuit with an **SPST** (single-pole, single-throw) switch connected in series with the holding contact.

In the closed position, the SPST switch offers no opposition to the flow of current and the circuit behaves the same as a standard three-wire circuit. The **normally open** push button is acting as a "start" or "run" button.

If the SPST switch is opened, then it has introduced an open in series with the **2-3 holding contacts**, effectively removing them from the circuit. Without the holding contact, the motor starter will only be energized as long as an operator is pressing the normally open pushbutton, which acts as a "jog" button in this position.

- The main advantage of this circuit is the ease of installation and the cheapness of equipment.
- The main disadvantage is that you must change the position of your selector switch to change the function of your button.

Jog circuit with a selector switch

A BCcampus element has been excluded from this version of the text. You can view it online here: https://pressbooks.bccampus.ca/basicmotorcontrol/?p=78 (https://pressbooks.bccampus.ca/basicmotorcontrol/?p=78#pb-interactive-content)

Dangerous Jog Circuit

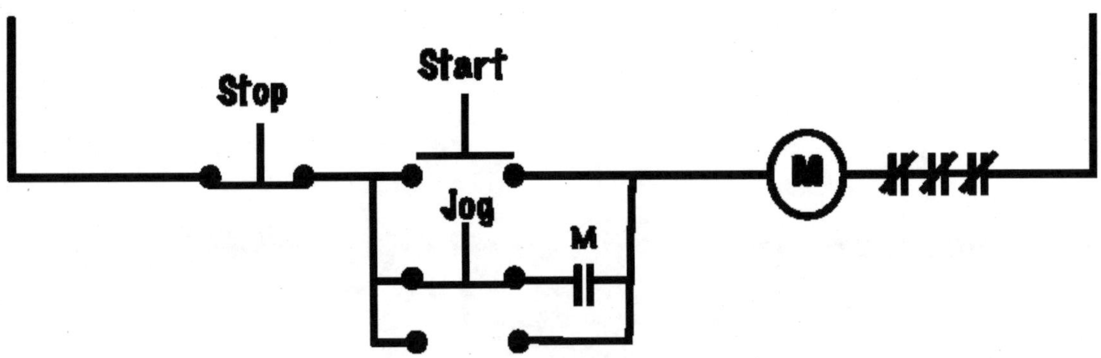

Dangerous jog control circuit

The next jog circuit we will look at uses a four-contact momentary pushbutton as our "Jog" button. This button has one set of **normally closed contacts** which are wired in series with the 2-3 holding contact, and one set of normally open contacts which are in series with only the stop button and the motor starter.

In normal operation, the stop and start buttons provide their standard functions in a three-wire circuit, as the 2-3 holding contact is in series with the normally closed contacts of the jog button. If the jog button is pressed, the normally closed contacts will open and the normally open contacts will close, providing a path for current to energize the motor starter.

When the motor starter is energized, all contacts associated with it will change their state, including the 2-3 holding contact, but because the jog button is being depressed the holding contact cannot maintain power to the starter. Once the jog button is released the motor comes to a stop.

This circuit is sometimes referred to as the "dangerous jog circuit" for reasons which may appear obvious now. If the jog button's normally closed contacts are able to return to their normal condition before the motor starter's **armature** has had a chance to drop out, then the coil will remain energized and the motor will continue to run. This is dangerous because if an operator pushes a jog button, expecting the motor to stop when they release the button, and the motor continues to run, it could introduce a hazard to a person caught off guard. In short, we never want machines to surprise people.

- This circuit has the advantage of being simple to install and has separate buttons dedicated to starting and jogging the motor.
- The main disadvantage is the hazard introduced by the quick return of the jog button to its normal state.

The bad jog circuit
(Contact racing)

A BCcampus element has been excluded from this version of the text. You can view it online here: https://pressbooks.bccampus.ca/basicmotorcontrol/?p=78 (https://pressbooks.bccampus.ca/basicmotorcontrol/?p=78#pb-interactive-content)

Jog Circuit with Control Relay

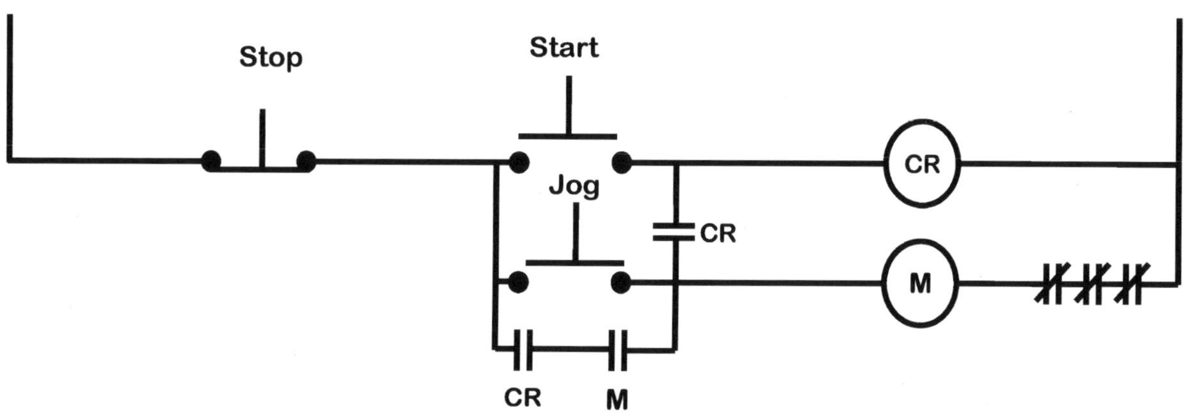

Jog circuit with control relay

A more sophisticated jogging circuit uses a **control relay** as shown in the figure above. Control relays behave just like motor starters but lack overload protection and power contacts. Control relays are loads that must be connected in **parallel** with the motor starter to ensure they get their rated **voltage**.

In any **schematic diagram**, the current must find its way from Line 1 to Line 2 and energize only one load along the way. Switches offer either infinite **resistance** when they are open or zero resistance when they are closed, so some device must limit the current to prevent short circuits. Notice that the current that passes through the control relay does not pass through the motor starter. This means that they will both get their rated value of voltage and pull in their armatures.

As a rule, we NEVER connect loads in series.

Under normal conditions, if the start button is pressed, current will be able to complete the circuit and energize the control relay. Once the relay is energized, the two normally open contacts associated with it will change their state and close. This will provide a path for current to energize the motor starter, closing the 2-3 holding contacts and running the motor.

The circuit will continue to operate as a standard three-wire circuit providing **low-voltage protection (LVP)** until either the stop button is pressed, or an **overload** occurs.

If the jog button is pressed while the motor is running, there will be no change to the circuit.

If, however, the jog button is pressed while the motor is not running, it will provide a path for current to energize the motor starter. The current will not be able to energize the control relay, and so once the jog button is released, the starter drops out and the motor comes to a stop.

- The advantage of this circuit lies in its safety and reliability. Having two separate buttons for "run" and "jog" functions increases ease of control for the operator.
- The disadvantage is the additional cost and installation time associated with the control relay and additional current drawn in the **control circuit** due to the second coil.

Jog with a control relay

A BCcampus element has been excluded from this version of the text. You can view it online here: https://pressbooks.bccampus.ca/basicmotorcontrol/?p=78 (https://pressbooks.bccampus.ca/basicmotorcontrol/?p=78#pb-interactive-content)

36.

Forward/Reverse Control Circuits

If a three-phase motor is to be driven in only one direction, and upon its initial energization it is found to be rotating opposite to what is desired, all that is needed is to interchange any two of the three line leads feeding the motor. This can be done at the **motor starter** or at the motor itself.

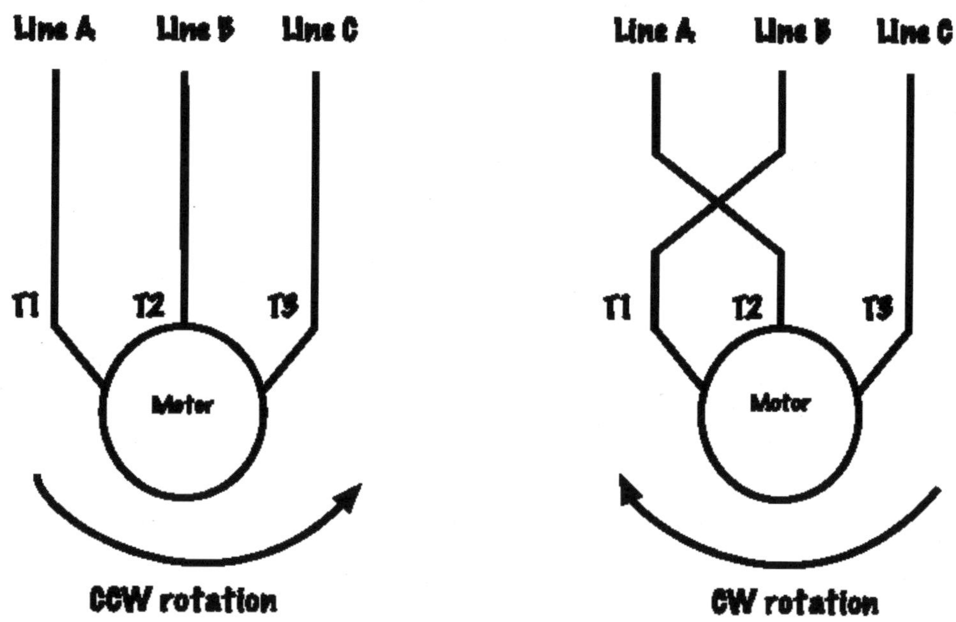

Three-phase motor rotation

Once two of the lines have been switched, the direction of the magnetic fields created in the motor will now cause the shaft to spin in the opposite direction. This is known as reversing the **phase rotation**.

Reversing Magnetic Starters

If a motor is to be driven in two directions, then it will require a Forward / Reverse motor starter, which has two three-pole horsepower-rated contactors rather than just one as in the conventional starter. Each of the two different motor starters powers the motor with a different phase rotation.

When the forward contactor is energized, power contacts connect line L1 to T1, line L2 to T2 and line L3 to T3 at the motor. When the reverse contactor is energized, the power contacts connect line L1 to T3, line L2 to T2 and line L3 to T1 at the motor.

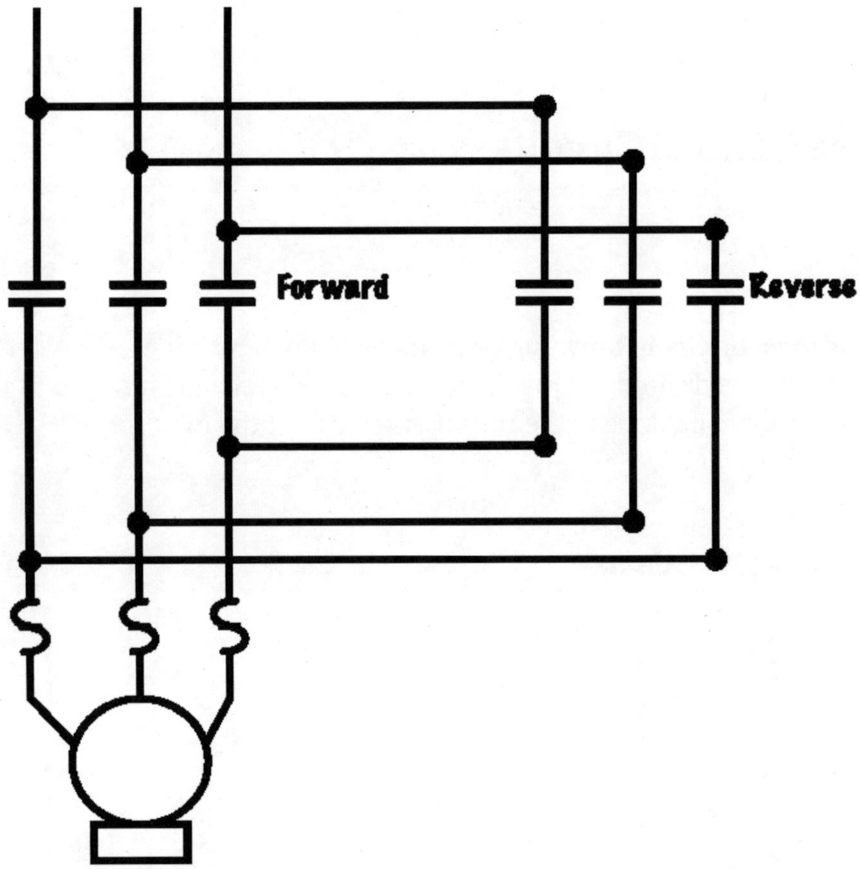

Forward/Reverse power circuit

Since the two motor starters control only one motor, only one set of overload relay heaters need be used. The return paths for both starter coils connect in **series** with the **normally closed contacts** of the **overload relay** so that if an overload occurs in either direction, the starter coils will be de-energized and the motor will come to a stop.

Note that the two contactors must be **electrically** and **mechanically interlocked** so that they cannot be energized simultaneously. If both starter coils became energized simultaneously, a short circuit will occur with potentially hazardous results.

Forward / reverse starters will come with two sets of normally open **auxiliary contacts** to act as holding contacts in each direction. They will also come with two sets of normally closed auxiliary contacts to act as electrical interlocks.

Mechanical Interlocks

Forward / reverse starters must never close their power contacts simultaneously. The best way to provide this is through electrical interlocks, which prevent the one coil from being energized if the other is engaged. A failure in electrical interlocking can cause both coils to be energized at the same time.

If both become energized, some form of mechanical interlock is required to prevent both **armatures** from pulling in. Represented on **schematic diagrams** as a dotted line between the two coils, a mechanical interlock is a physical barrier that is pushed into the path of one coil's armature by the movement of the adjacent coil. This means that even if both coils are energized, only one armature will be able to pull in fully. The coil that is prevented from pulling in will make a terrible chattering sound as it tries to complete the magnetic circuit.

Mechanical interlocks should be relied on as a last resort for protection.

Electrical Interlocks

Electrical interlocking is accomplished by installing the normally closed contact of one direction's coil in series with the opposite direction's coil, and vice versa. This ensures that when the forward coil is energized, pushing the reverse **pushbutton** will not energize the reverse coil. The same situation is in effect when the reverse coil is energized. In both situations the stop button will need to be pressed to de-energize the running coil and return all its auxiliary contacts back to their original state. Then the opposite direction coil can be engaged.

Reversing Control Circuit

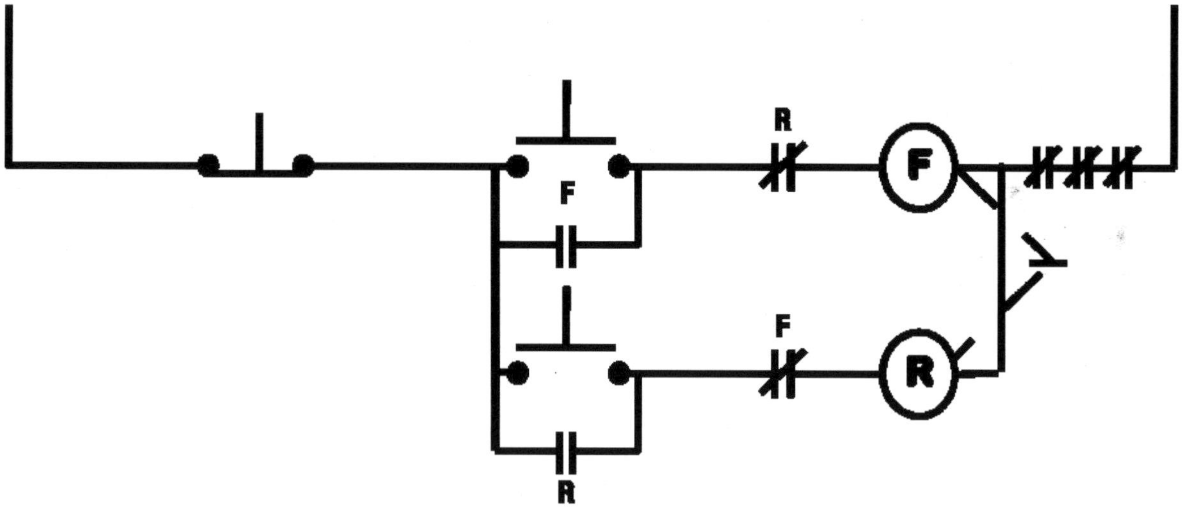

Forward/Reverse control circuit

When designing the control schematic for forward / reverse circuits, we start with the standard **three-wire circuit**, add a second normally open pushbutton, and add a holding contact branch for the second coil. A single stop button is sufficient to disable the motor in both directions.

The two coils are mechanically interlocked and the normally closed instantaneous contacts provide electrical interlocking.

If the forward pushbutton is pressed, as long as the reverse coil is not engaged, current will find a path through the normally closed reverse contact and energize the forward coil, causing all **contacts** associated with that coil to change their state. The **2-3 holding contact** will close and the normally closed electrical interlock will open. If the reverse pushbutton is pressed while the forward coil is engaged, current will not be able to get past the forward normally closed contact, and nothing will happen.

In order to send the motor in the reverse direction, the forward coil must be de-energized. To do this, the stop button must be pressed, then the reverse pushbutton will be able to energize the reverse coil.

Regardless of the direction the motor is spinning in, this circuit will operate as a standard three-wire circuit providing **low-voltage protection (LVP)** until either the stop button is pressed, or an **overload** occurs.

Pushbutton Interlocks

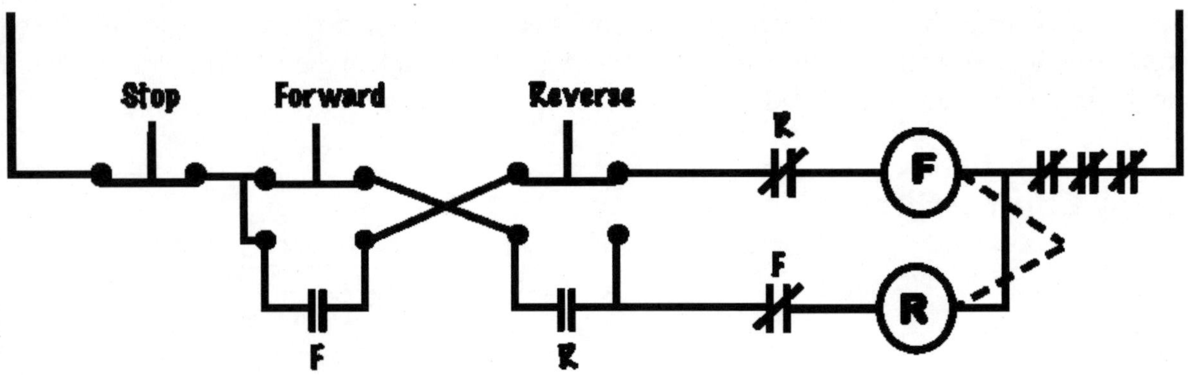

Forward/Reverse pushbutton interlock

Pushbutton interlocking requires the use of four-contact momentary push buttons with each pushbutton having a set of normally open and normally closed contacts.

To achieve pushbutton interlocking, simply wire the normally closed contacts of one pushbutton in series with the normally open contacts of the other pushbutton, and the holding contacts will be connected in **parallel** with the appropriate button's normally open contacts.

This circuit still requires the installation of electrical interlocks.

Pushbutton interlocking doesn't require the motor coils to be disengaged before reversing direction because the normally closed forward contacts are in series with the normally open reverse contacts, and vice-versa. Pushing one button simultaneously disengages one coil while starting the other. This sudden reversal **(plugging)** can be hard on the motor, but if quick reversal of the motor is required, this circuit can be a solution.

Forward and Reverse Circuit

A BCcampus element has been excluded from this version of the text. You can view it online here: https://pressbooks.bccampus.ca/basicmotorcontrol/?p=80 (https://pressbooks.bccampus.ca/basicmotorcontrol/?p=80#pb-interactive-content)

37.

Plugging with a Zero-Speed Switch

When a motor is de-energized, the shaft continues to spin until inertia and friction cause it to slow down and come to a halt. If this condition is undesirable (for example, if we want the motor to stop instantly), we can use a **plugging circuit**, which requires a forward / reverse motor starter and a **zero-speed switch**.

Although the motor operates in one direction only, a reversing magnetic starter is needed.

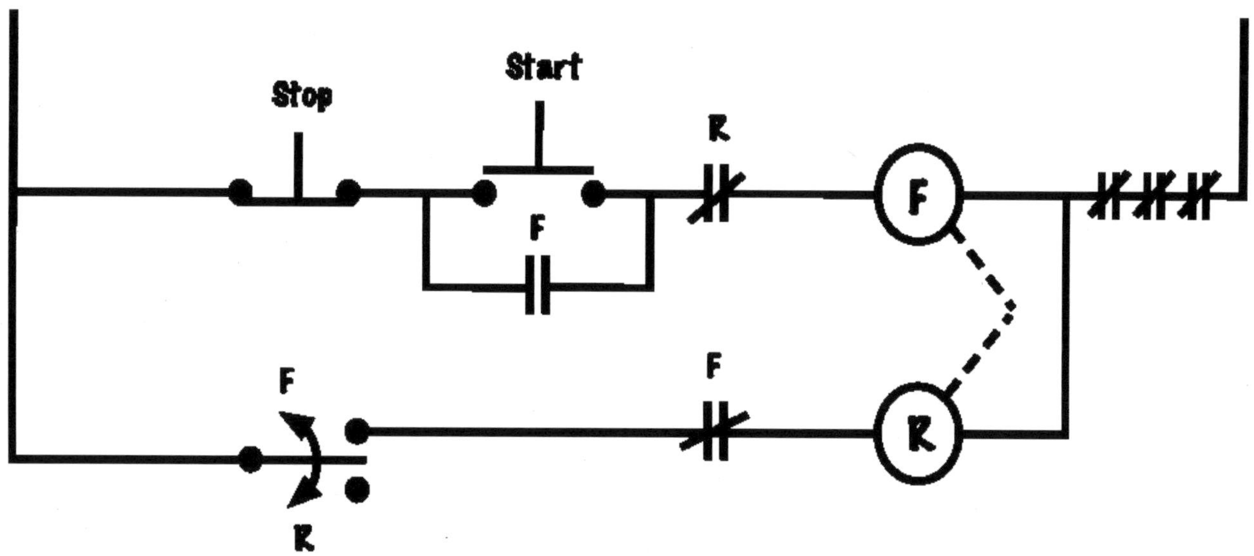

A zero-speed switch is a switch that is activated by centrifugal forces and is attached either directly or via a belt to the shaft of the motor. To provide **plugging** action, the zero-speed switch must be wired **normally open** so that any motion of the motor shaft will cause the **contacts** to change their state and close.

To **plug** a motor to a stop, the circuit momentarily reverses the motor while it is rotating in the forward direction, but then de-energizes it before it can build up any momentum in the reverse direction.

Under normal operating conditions, the motor is started in the forward direction by pressing the start button. This energizes the forward coil and closes the normally open **2-3 holding contact** in **parallel** with the start button as well as opening the normally open **electrical interlocks** in **series** with the reverse coil.

The motor will continue to run and the circuit will operate as a standard **three-wire circuit** providing **low-voltage protection (LVP)** until either the stop button is pressed, or an **overload** occurs.

While the shaft of the motor is spinning, the contacts of the zero-speed switch will change their state, and the forward contact will close. This contact is wired in series with the reverse coil, but cannot energize it so long as the electrical interlock remains open.

When the stop button is pushed, the forward coil is de-energized, dropping out its **armature**. This causes all the contacts associated with that coil to return to their normal condition, including the normally closed electrical interlock.

Since the shaft of the motor is still spinning in the forward direction due to inertia, the normally open zero-speed switch contacts are being held closed. This means that the instant the stop button is pressed, the reverse coil gets energized, momentarily sending the motor in the reverse direction and cancelling out all of its forward momentum, thus bringing the motor to a halt much quicker.

The instant that the shaft of the motor begins to move in the reverse direction, the held-closed forward contacts of the zero-speed switch open and the reverse coil is de-energized before the motor can build up any momentum in the reverse direction.

The motor has now been plugged to stop. In practice, this all takes place in a fraction of a second.

Plugging with a zero
Speed switch

A BCcampus element has been excluded from this version of the text. You can view it online here: https://pressbooks.bccampus.ca/basicmotorcontrol/?p=97 (https://pressbooks.bccampus.ca/basicmotorcontrol/?p=97#pb-interactive-content)

38.

Anti-Plugging With Zero-Speed Switch

If a forward-reverse circuit is not designed or rated to withstand the severe over**currents** and torques associated with sudden reversals in direction, a **zero-speed switch** wired **normally closed** can provide this function.

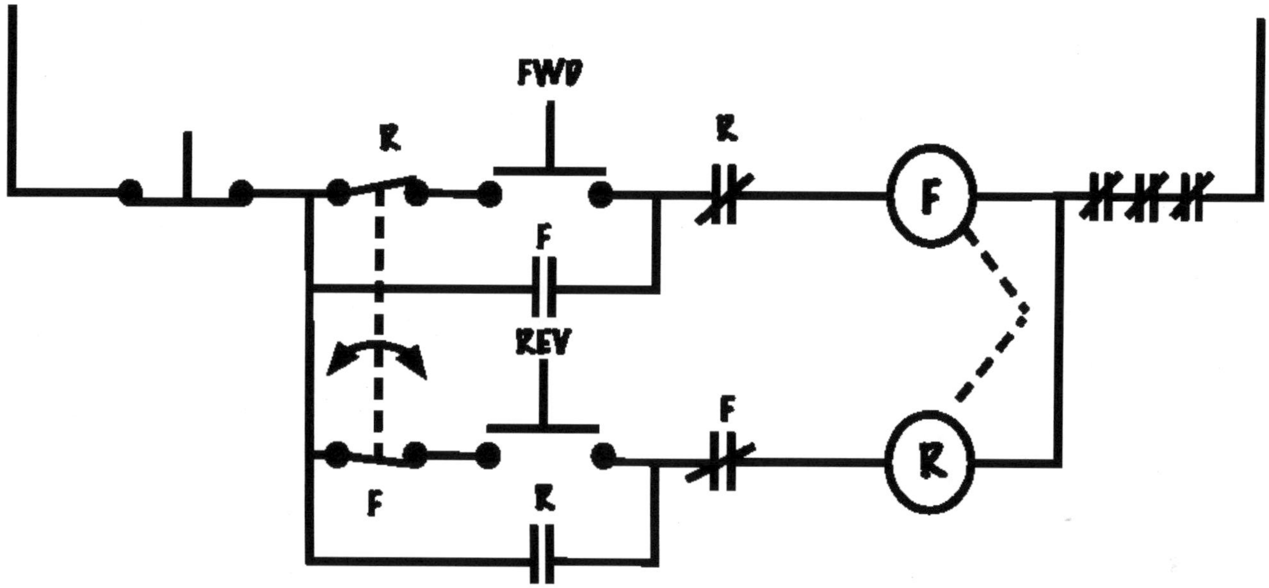

Once the zero-speed switch is attached to the shaft of the motor, its normally closed contacts are then wired in **series** with the opposite directions coil. In this manner, they act similar to the **electrical interlocks** but provide an extra delay function while the shaft of the motor still spins.

This means that even after the stop button has been pressed, the reverse coil cannot be engaged until the motor has stopped rotating in the forward direction, and vice-versa. Once the motor has come to a rest the zero-speed switch contacts will close and the motor can then be engaged in the opposite direction.

Anti-plugging with
A zero speed switch

A BCcampus element has been excluded from this version of the text. You can view it online here: https://pressbooks.bccampus.ca/basicmotorcontrol/?p=99 (https://pressbooks.bccampus.ca/basicmotorcontrol/?p=99#pb-interactive-content)

39.

Timer Circuits Overview

So far all the circuits we have been looking at have all used instantaneous contacts. Both **motor starters** and **control relays** have **contacts** that change their state as soon as the coil is energized.

If a circuit requires some sort of delay in timing action, perhaps for sequence control or safety purposes, then timing relays can be used to provide this function. For example, a conveyor belt system might have several motors installed in it. Starting each motor at the same time could induce a very large **inrush current** on the system. One way to minimize the effects of these inrush currents is to stagger the starting of the motors with on-delay timers.

If this same conveyor belt system was designed so that each segment fed onto the next and at the end of the work day each conveyor segment had to be cleared, using off-delay timers would allow the circuit to be arranged so that after pressing a single stop button, each of the motors continued to run for a limited time until all product was cleared off them.

Timers can also be used to provide both **plugging** and **anti-plugging** for certain circuits. Indeed the range of application for timers is nearly limitless.

40.

Sequence Control: On-Delay

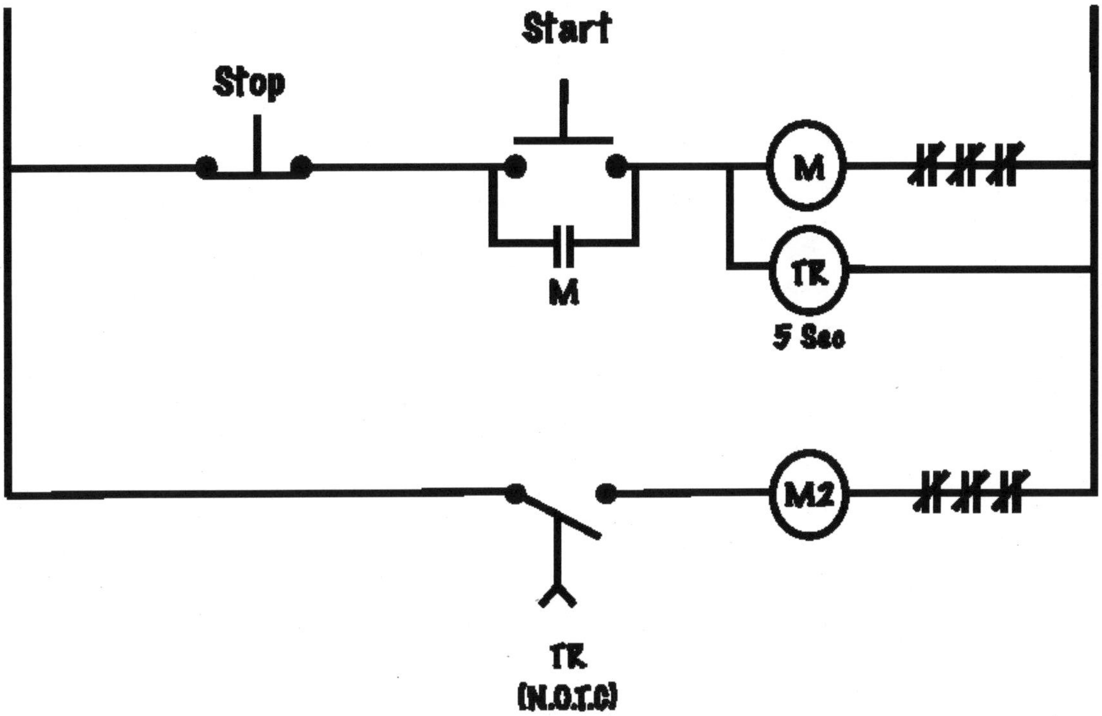

Definite sequence on-delay

The above circuit shows a standard **three-wire circuit** for a single **motor starter** M1. In **parallel** with M1 is a **time-delay relay (TR)** whose normally open, timed to close (NOTC) contacts identify it as an on-delay timer. These timed contacts are in **series** with motor starter M2. The above switching arrangement will allow two motors to be controlled from a single **pushbutton** station and will guarantee that M2 energizes five seconds after M1 is energized. Once both motor starters are engaged, the motors will continue to run until the stop button is pressed.

If an overload occurs on motor M2, then only that motor will stop, but if motor M1 develops an overload and its **OLR contacts** open, then the coil will de-energize and its **2-3 holding contact** will open, disconnecting the timer coil from the source of supply. Once the on-delay coil is de-energized, its contacts will revert to their original state instantly, so motor M2 will also be de-energized.

If more than two motors are to be sequenced in this fashion, we would simply scale up this switching arrangement by connecting additional timer relays in parallel with each motor starter so that as one motor starts, it also begins a countdown for the next one in sequence.

Time Delay Relays
(On Delay)

A BCcampus element has been excluded from this version of the text. You can view it online here: https://pressbooks.bccampus.ca/basicmotorcontrol/?p=91 (https://pressbooks.bccampus.ca/basicmotorcontrol/?p=91#pb-interactive-content)

41.

Sequence Control: Off-Delay

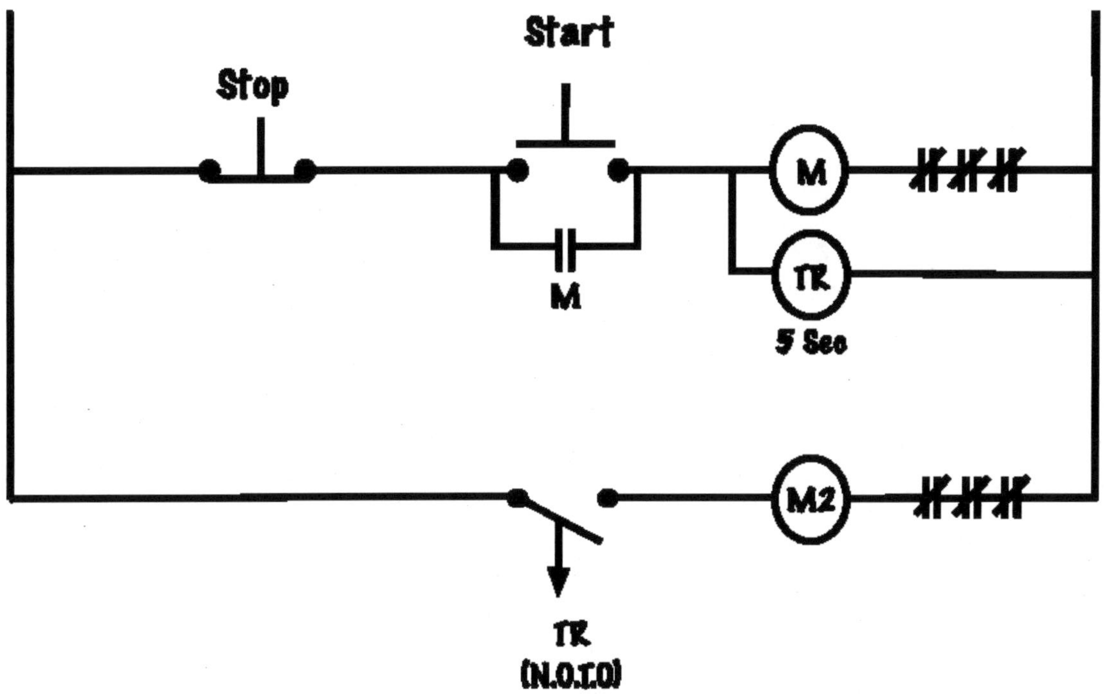

Definite sequence control off-delay

The above circuit shows a standard **three-wire circuit** for a single-motor starter M1. In **parallel** with M1 is a **time-delay relay (TR)** who's normally open, timed to open (NOTO) contacts identify it as an off-delay timer. These timed contacts are in **series** with motor starter M2.

The above switching arrangement will allow two motors to be controlled from a single **pushbutton** station. If the start button is pressed both motors M1 and M2 will start instantly. This is because the **normally open contacts** associated with the off-delay coil will change their state instantly when the coil is energized.

Once both motor starters are engaged, the motors will continue to run until the stop button is pressed. Once pressed, the M1 contactor and the off-delay timer will be de-energized, and their contacts will revert to their original state.

For the **motor starter,** this will happen instantly, but the timed contacts associated with the timer coil will have a delay of five seconds before they open, during which time motor M2 will continue to run. It is important to note that even though the timer coil has been disconnected from its source of power, it still performs its timing function. It does not need external energy to delay its contacts, that energy is stored in the timer, usually as compressed air or spring tension.

If an **overload** occurs on motor M2, only that motor will stop, but if motor M1 develops an overload and its **OLR contacts** open, then the coil will de-energize and its 2-3 holding contact will open, disconnecting the timer coil from the source of supply. Once the off-delay coil is de-energized, its contacts will delay for five seconds, then revert to their original state, so motor M2 will continue to run for five seconds after motor M1 has stopped due to an overload.

If more than two motors are to be sequenced in this fashion, we would simply scale up this switching arrangement by connecting additional timer relays in parallel with each motor starter so that each motor stops one after the other in the prescribed sequence.

Time Delay Relay
(Off delay)

A BCcampus element has been excluded from this version of the text. You can view it online here: https://pressbooks.bccampus.ca/basicmotorcontrol/?p=93 (https://pressbooks.bccampus.ca/basicmotorcontrol/?p=93#pb-interactive-content)

42.

Anti-Plugging with Timers

A forward/reverse circuit can use two off-delay timers to provide anti-plugging action in an **anti-plugging circuit**.

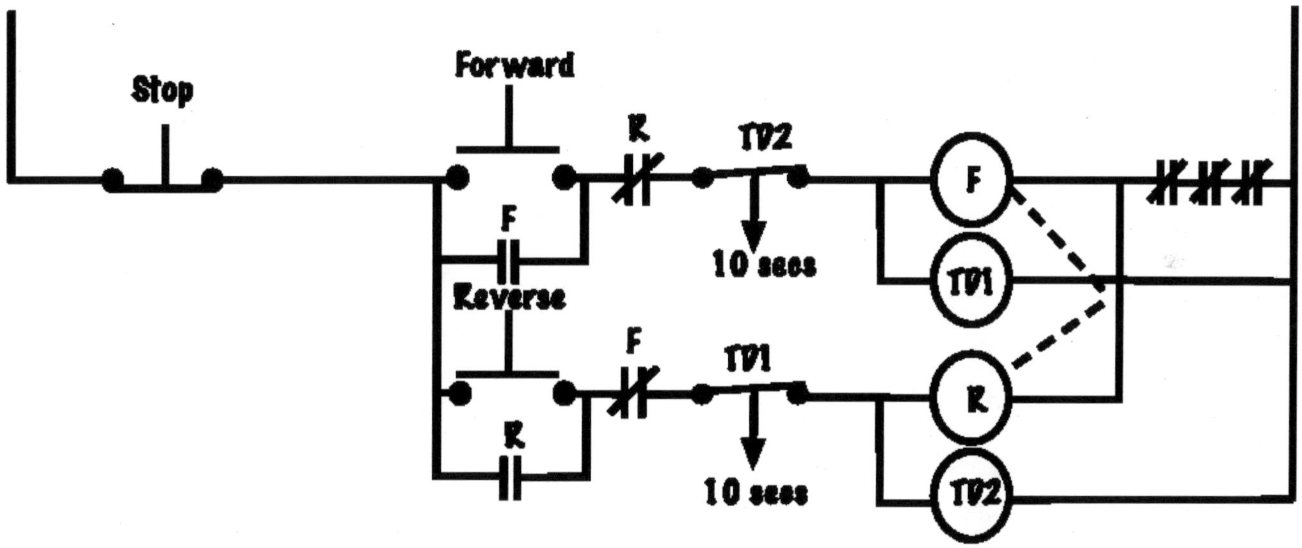

This circuit utilizes two off-delay coils to prevent the forward and reverse coils from being energized until ten seconds after the stop button has been pressed. This will give the shaft of the motor time to wind down to a halt and provide an anti-plugging feature for the circuit. To achieve this, the normal **control circuit** for a reversing magnetic starter is modified slightly by connecting an off-delay timer in **parallel** with each of the contactor coils.

Similar to how the normally closed **electrical interlocks** are connected in the circuit, the normally closed timed to close (NCTC) contacts of the timer connected in parallel with the forward coil are connected in **series** with the reverse motor starter, and vice versa.

If, while the motor is not energized, the forward **pushbutton** is closed, TD1 and the forward starter will energize, and the motor will begin to spin in the forward direction.

If, while the motor is not energized, the reverse pushbutton is closed, TD2 and the reverse starter will energize, and the motor will begin to spin in the reverse direction.

If the motor is energized in the forward direction, the circuit arrangement ensures that for ten seconds after the stop button has been pressed, the motor cannot be sent in the reverse direction as the timed contacts will remain open. Nothing however prevents the motor from being sent in the forward direction again, as the timed contacts for TD2 remain closed.

If the motor is sent in the reverse direction, then stopped, TD2 will ensure that the operator must wait ten seconds before the motor can be sent in the forward direction.

Under normal operation, this circuit will provide **low-voltage protection (LVP)**, and the motor will run in either direction until either the stop button is pressed, or an **overload** occurs.

43.

Self Test 5

An interactive or media element has been excluded from this version of the text. You can view it online here: https://pressbooks.bccampus.ca/basicmotorcontrol/?p=521 (https://pressbooks.bccampus.ca/basicmotorcontrol/?p=521#pb-interactive-content)

If using the print, PDF, or eBook copy of this book, navigate to the above link to complete the quiz. However, the quiz questions are also provided at the end of the book for offline use: Offline Copies of Chapter Quizzes

VI

Troubleshooting

Learning Objectives

After studying this section, students should be able to:

- Identify common problems associated with electrical circuits.
- Use an ohmmeter for continuity checks in a circuit.
- Use a voltmeter for troubleshooting purposes.
- Describe how to use a voltmeter to check the status of fuses in a three-phase disconnect.

44.

Common Issues

This chapter will be focusing on troubleshooting common problems found in the **power** and **control circuits**. Troubleshooting motors and the mechanical systems they drive is an extensive topic beyond the scope of this text.

A motor may not start for a variety of reasons. Knowing what to check for and where to begin is a valuable skill that can reduce downtime when a fault occurs.

Usually, there are two fundamental steps in troubleshooting. The first is a cursory visual inspection, followed by a more in-depth examination with testing equipment and referencing several diagrams

Motor control circuits have several protective devices built into them to protect wires and equipment. It is always a good idea to start your visual inspection by checking overload and overcurrent devices. A visual inspection sometimes quickly locates a fault or identifies situations requiring maintenance. For example, you might see a tripped indicator on a **circuit breaker**, **switch** or **overload relay**. Resetting the device might be all the circuit needs to resume working.

If an **overload** or **overcurrent** situation has occurred, it is advisable to find the source of this fault. A visual inspection can identify burnt or heat-discoloured components, loose wires in terminal blocks, or mechanical parts that bind together or cause excessive friction.

When there is no obvious cause for the problem(s), it is often necessary to analyze **circuit diagrams** and take measurements of **voltage** and **resistance** at key points of the circuit. To do this, we use **voltmeters** and **ohmmeters**. A firm understanding of their use and limitations is necessary for their safe operation in circuits.

45.

Using Meters for Troubleshooting

Voltmeter

A **voltmeter** is designed to be used on a live circuit. When working on live circuits, all safety precautions and PPE (personal protective equipment) requirements should be adhered to.

A voltmeter is essentially a very high value of **resistance**. When the voltmeter leads are connected to two points in a circuit, that high value of resistance is connected in **parallel** with that point and, due to **Kirchhoff's Voltage Law**, experiences the same value of **voltage**. By measuring how much current flows across the meter's internal resistance, it can calculate the voltage value.

Voltmeters typically read zero volts when measuring between points of equal potential. Closed **switches** and connected wires are examples of components that, when energized, are of equal potential.

Voltmeters typically read line voltage when measuring between points of different potential. In a **control circuit** this is line voltage. In the **power circuit**, this is a phase-to-phase voltage, which is different from their phase-to-ground voltage.

Open switches and **contacts** are examples of components that a voltmeter would measure line voltage across.

When current flows through a circuit, it drops off voltage proportional to the resistance of the device it flows through. Since closed switches and contacts offer nearly zero resistance, there will be no voltage drop measured across them. In every branch of a control schematic, there must be only one load to limit the current, and this device will have line voltage dropped across it when the circuit is operating. The coils of **motor starters**, **control relays**, and **timer relays** are examples of loads that when energized would have full line voltage dropped across them. **Pilot lights** are another example of resistive loads that would experience full line voltage.

Ohmmeter

Disconnect circuit from power supply first! Before using an **ohmmeter** in a circuit, use a voltmeter to confirm that power is off and that there is zero potential difference between the two points you wish to measure.

An ohmmeter works by using an internal voltage source to push a small DC current through its leads. By measuring the value of current, it can display a calculated value of ohmic resistance. Because it has an internal voltage source, ohmmeters cannot ever be connected in live circuits as they could cause damage to equipment or injury to the operator.

When using an ohmmeter in a control circuit, there are three typical readings you can get:

1. **Close-to-zero ohms:** This means that the leads of the ohmmeter are connected across two points that are **electrically common**. The two terminals of a closed switch or contact would give an ohmmeter reading close to zero **ohms**.
2. **Very-high-to-infinite ohms:** An open in the circuit would allow zero current to flow and so would read as infinite ohms. The terminals of open switches and contacts would give very high ohmic readings.
3. **Some ohms:** An ohmmeter measuring across a load such as a pilot light would read a very high (approximately megaohms), but not infinite, value of ohms. This is one way of confirming that a pilot light is in good working condition. The terminals of a coil should have continuity and a low (approximately tens to hundreds of ohms), but not zero value of resistance. If a coil is shorted and reads zero ohms, it needs to be replaced.

When using an ohmmeter to test **fuses**, first confirm they are removed from the circuit. If a fuse is in good condition, it should give a close-to-zero reading of resistance. If the fuse has blown due to a fault, then it should behave as an open and give an infinite resistance reading.

A BCcampus element has been excluded from this version of the text. You can view it online here: https://pressbooks.bccampus.ca/basicmotorcontrol/?p=103 (https://pressbooks.bccampus.ca/basicmotorcontrol/?p=103#pb-interactive-content)

46.

Check for Incoming Power

Whenever equipment or circuits are not working, always start by confirming that there is **voltage** available. If there is voltage at the source of supply, then you have confirmed that the problem is downstream from that point.

If the main **overcurrent** protection is provided by a **circuit breaker**, check that the handle is in the "On" position, and not the "Off" or "Tripped" position. A circuit breaker indicates that it has tripped when the handle is in the centre position. To reset, push the handle fully off, then on. If the handle "trips" again suddenly, then there is likely a short circuit somewhere bypassing a load.

If the main overcurrent protection is provided by a fusible disconnect, the status of the **fuses** will determine if voltage is present. There are two tests that can confirm if a fuse link is intact: an **ohmmeter** continuity test and the **crisscross voltmeter method**.

47.

Crisscross Voltmeter Method

When testing **fuses** in the **power circuit** supplying a three-phase motor, we use the **crisscross voltmeter method**.

The **horsepower-rated motor contacts** must be open and the three-phase disconnect must be closed in order to obtain proper readings.

There are three sets of measurements to perform on the line side of fuses to make sure that **voltage** is available. Measure across each pair of line terminals (L1–L2; L2–L3; L3–L1). In the diagram below, that would mean using a **voltmeter** to check between points 1-3; 3-5; 5-1. If any of these tests give a reading other than the phase-to-phase voltage, then check the incoming voltage upstream. If all three readings give a phase-to-phase voltage, then we know that voltage is present in the power circuit up to the fuses. The next test will confirm if the fuses are in good shape.

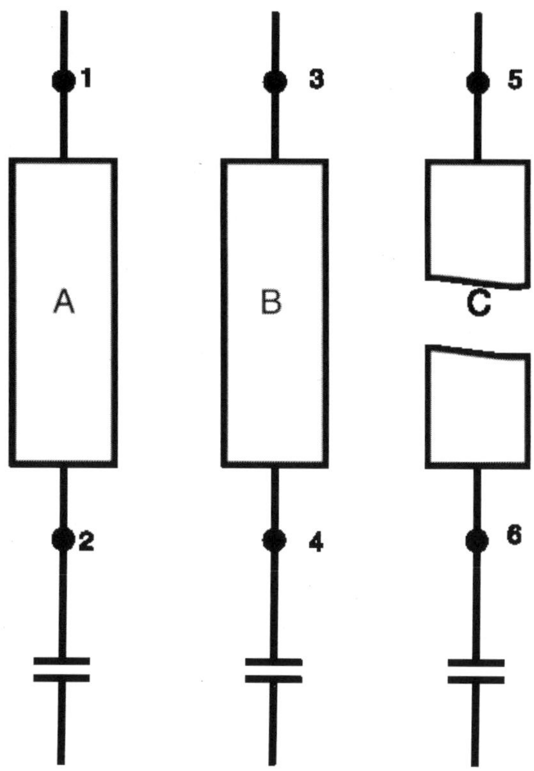

Three phase fuses, fuse C is blown

In the diagram above, all three readings give us a line-to-line voltage. To check the condition of the fuses, we measure from the line side of one fuse to the load side of another fuse.

Using the diagram above, that would mean using a voltmeter to check between points 1-4; 3-6; 5-2. The readings we get back are:

- 1-4 = line voltage, therefor Fuse B is good
- 3-6 = zero volts, therefor Fuse C is blown
- 5-2 = line voltage, therefor Fuse A is good

Because fuses A and B are in good condition, there is essentially no potential difference between points 1 and 2 and between points 3 and 4 respectively, and that is why the voltmeter reads line voltage on both sides of the fuse.

With the power contacts open and fuse C blown, the voltmeter lead connected to point 6 is completely isolated from any other part of the circuit and so experiences zero potential difference.

This method is called the crisscross voltmeter method because it is never necessary to check voltage through a fuse. If the fuse is in good condition as fuses A and B are, then we are measuring from points of equal potential, and if the fuse is blown, then our second voltmeter lead is isolated from the circuit, again giving us no potential difference. No useful information can be obtained from these measurements.

Criss Cross method
Of testing fuses

A BCcampus element has been excluded from this version of the text. You can view it online here: https://pressbooks.bccampus.ca/basicmotorcontrol/?p=111 (https://pressbooks.bccampus.ca/basicmotorcontrol/?p=111#pb-interactive-content)

48.

Voltage Drops in Control Circuits

An example of voltage drops in a **control circuit** will illustrate the use of **voltmeters** for troubleshooting.

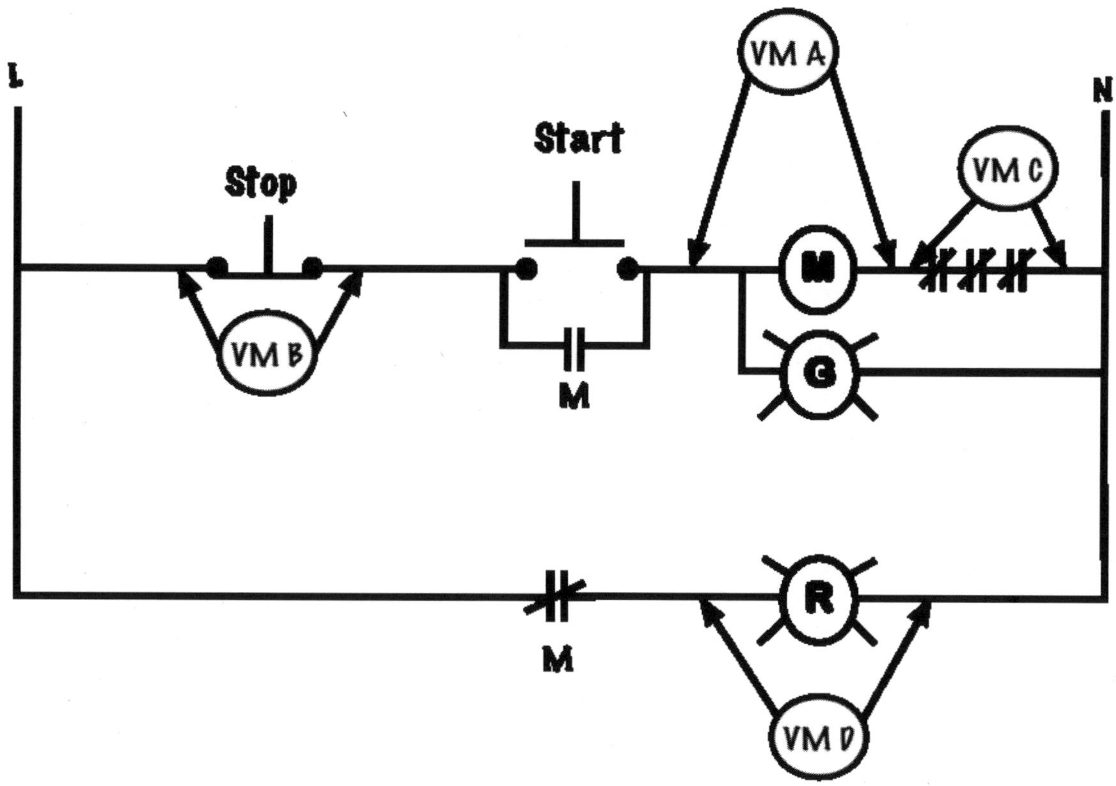

Troubleshooting control schematic

Referring to the diagram above and assuming a line **voltage** of 120V, if the circuit is operating normally we can determine what all the meters would be reading.

If someone were to press the start button, then only voltmeter A would read line voltage, as it is in **parallel** with the two loads.

Voltmeter B will measure zero volts across the closed contacts of the stop button.

If an **overload** occurs and the **OLR contacts** open, voltmeter C will read line voltage while the start button is being pressed, and nothing when it is released.

Assuming the motor is not running, only voltmeter D would give a reading of line voltage. If the contactor is not engaged, then the **normally closed contact** will remain closed and the red **pilot light** will be on.

49.

Self Test 6

An interactive or media element has been excluded from this version of the text. You can view it online here: https://pressbooks.bccampus.ca/basicmotorcontrol/?p=523 (https://pressbooks.bccampus.ca/basicmotorcontrol/?p=523#pb-interactive-content)

If using the print, PDF, or eBook copy of this book, navigate to the above link to complete the quiz. However, the quiz questions are also provided at the end of the book for offline use: Offline Copies of Chapter Quizzes

Offline Copies of Chapter Quizzes

Terms and Definitions – Self Test 1

Questions

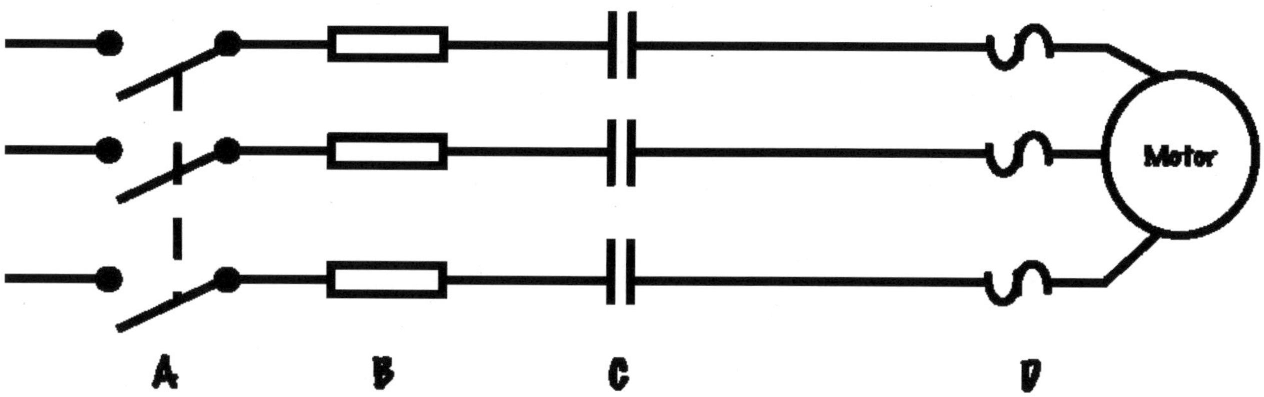

Using the above diagram, answer questions 1 to 5:

1. Which letter identifies the component that normally provides overcurrent protection for the motor branch circuit?
2. Which letter identifies the component that normally provides overload protection for the motor branch circuit?
3. When energized by the control circuit, which letter identifies the component that provides normal starting and stopping actions?
4. Between which two letters is the control circuit power normally taken from?
5. Which letter(s) identify devices that must be horsepower rated?

6. A melting alloy type of thermal overload is called a:
   A. Solder-pot relay
   B. Dash-pot relay
   C. Thermal relay
   D. Bimetallic relay
7. The part of a manual AC motor starter which senses motor overload current is the:
   A. High-limit switch

B. Solder-pot assembly

C. Overload contact

D. Heater element

8. If automatic starting after a power failure represents a safety hazard for a motor drive, it should be equipped with:

A. Low voltage release

B. A flashing red light

C. Low-voltage protection

D. A warning sign

9. An overload relay that uses a strip of dissimilar metals is called a _______ relay

A. Melting alloy

B. Thermistor

C. Solder pot

D. Bimetallic

10. Referring to the drawing, assuming that fuse C is blown, which pair of points will result in a reading of zero volts when the voltmeter is connected across them?

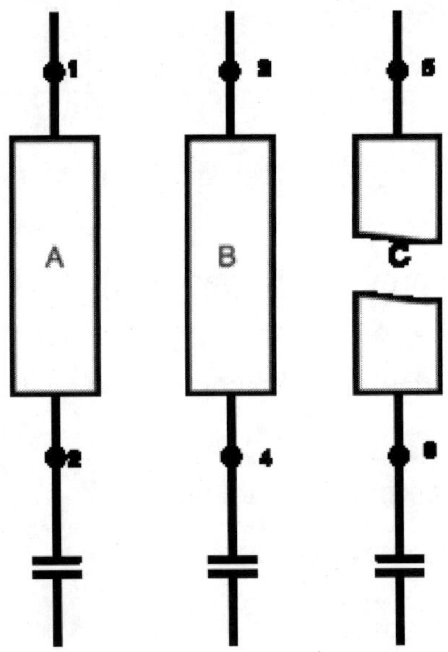

11. To connect a motor solely for jogging operation, the control circuit will:

A. Use holding contacts

B. Connect holding contacts in series with the start button

C. Connect holding contacts in parallel with the start button

D. Not use holding contacts

12. Timed relays are the best method to provide plugging for a motor. True or false?

13. The electrical interlocks in a reversing magnetic starter are:
    A. Controlled by the timing relays necessary for circuit operation
    B. In parallel with the forward and reverse pushbuttons
    C. Normally closed contacts
    D. Normally open contacts
14. Overload protection for motors is achieved by connecting:
    A. Thermal sensing elements in series with the motor
    B. A reset relay in series with the motor
    C. Cartridge fuses in series with the motor
    D. A set of N.C contacts in series with the motor
15. With a three-wire control circuit, when power is restored after a low voltage situation:
    A. An operator must restart the motor
    B. The motor will polyphase
    C. The motor will automatically restart
    D. The motor will automatically restart after a time delay

Answers

1. B
2. D
3. C
4. B and C
5. A and C
6. A
7. D
8. C
9. A
10. 3 and 6
11. D
12. False
13. C
14. A
15. A

Pilot Devices – Self Test 2

Questions

1. In a motor control circuit with multiple start/start pushbutton stations, the stop buttons would be connected in ______ and the start buttons would be connected in _______.
2. To troubleshoot an electrical control circuit, the best drawing to be used would be:
    A. Schematic
    B. Wiring

C. Pictorial

D. RIser

3. A double contact pushbutton is illustrated below. When connected as a start button in the control circuit of a magnetic starter, which pair of terminals are normally utilized?

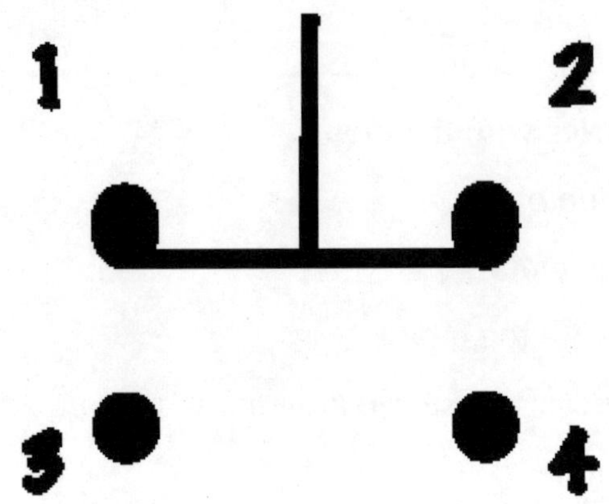

4. Timed relays are the best method to provide plugging for a motor. True or false?

Answers

1. Series, parallel
2. A
3. 3 and 4
4. False

Motor Starters and Contactors – Self Test 3

Questions

1. The minimum number of control circuit conductors to a stop/forward/reverse pushbutton station that provides low-voltage protection for a three-phase reversible motor is:

   A. 2 wires

   B. 3 wires

   C. 4 wires

   D. 6 wires

2. Which of the following type of motor starter does not normally provide running protection for the motor?

   A. A magnetic starter

B. A pushbutton starter

C. A toggle-switch starter

D. A drum-switch controller

3. If the circuit in the drawing was operating normally and an overload occurred, then:

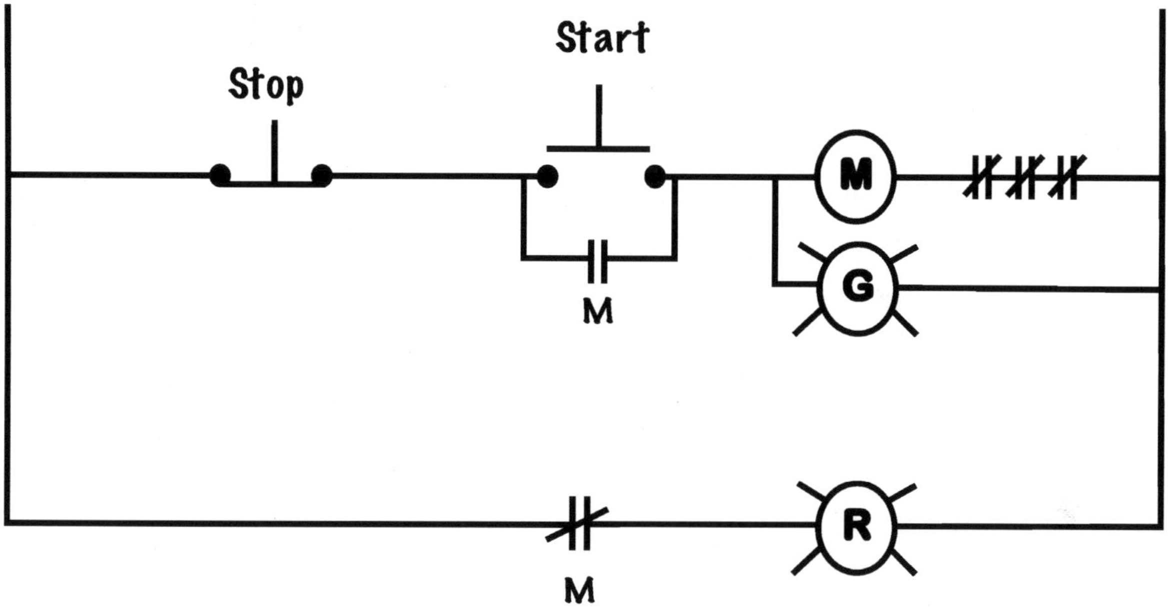

A. Both lights would come on

B. The green light would come on, and the red light would go out

C. Both lights would go out

D. The red light would come on, and the green light would go out

4. With a three-wire control circuit, when power is restored after a low voltage situation:

A. The motor will automatically restart after a time delay

B. The motor will automatically restart

C. An operator must restart the motor

D. The motor will polyphase

5. A melting alloy type of thermal overload is called a:

A. Solder-pot relay

B. Thermal relay

C. Bimetallic relay

D. Dash-pot relay

6. A starter rated at 10HP, 600 V, if used on a 120V motor would most likely be rated at:

A. 2 HP

B. 3 HP

C. 10 HP

D. 2.5 HP

7. A N.C.T.C contact will be open immediately after the coil is de-energized. True or false?

8. An overload relay that uses a strip of dissimilar metals is called a _______ relay

A. Thermistor

B. Melting alloy

C. Solder pot

D. Bimetallic

9. A shading coil is not required for a DC voltage coil. True or false?

10. Overload protection for motors is achieved by connecting:

A. A set of N.C contacts in series with the motor

B. Thermal sensing elements in series with the motor

C. Cartridge fuses in series with the motor

D. A reset relay in series with the motor

11. The part of a manual AC motor starter which senses motor overload current is the:

A. Overload contact

B. Solder-pot assembly

C. Heater element

D. High-limit switch

12. The essential difference between a magnetic motor starter and a magnetic contactor is that the contactor does not contain:

A. Shading coils

B. Holding contacts

C. Overload relay protection

D. Horsepower-rated contacts

13. The rotation of an AC three-phase induction motor may be reversed by interchanging:

A. The control circuit

B. The forward/reverse interlocks

C. The forward reverse coils

D. Any two power lines

14. If an AC magnetic contactor with a 480V coil was energized with 120V, then most likely the:

A. Overload relay would trip

B. Contactor would not pick up

C. Control circuit fuses would blow

D. Coil would overheat during normal operation

15. In a control schematic, a dotted line between two coils generally means that the two coils:

    A. Are operated together

    B. Are mechanically interlocked

    C. Are electrically interlocked

    D. Have a common set of contacts

16. During normal operation, a loud chattering noise is heard from the enclosure of an AC magnetic starter. What is the most likely cause?

    A. Broken shading coil

    B. Open in the seal-in circuit

    C. Rust on pole faces

    D. Power contact is not making good contact due to poor pressure

17. The purpose of electrical interlocking in a three-phase reversing AC magnetic motor starter is to:

    A. Ensure that the forward direction of rotation is selected first

    B. Allow both coils to be energized together

    C. Prevent both coils from being energized simultaneously

    D. Maintain the coil circuit after the stop button is released

18. A double-contact pushbutton is illustrated below. When connected as a start button in the control circuit of a magnetic starter, which pair of terminals are normally used?

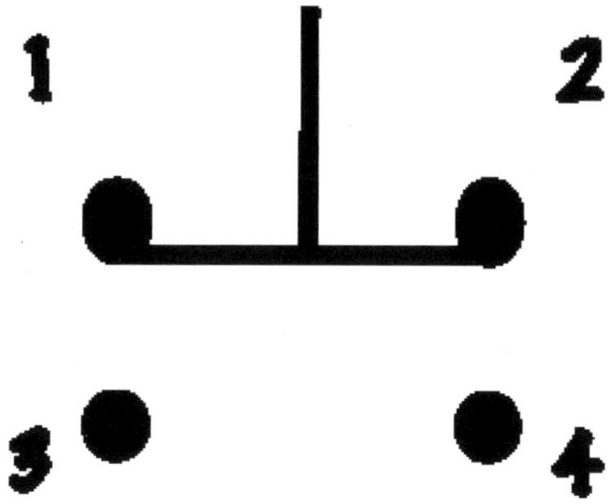

19. A time-delay relay includes both timed and instantaneous contacts. True or false?

20. Fill in the missing words. The electrical interlocks on a reversing starter are normally _______ contacts

21. A N.O.T.O contact will be closed while the coil of the relay is energized. True or false?

22. A N.C.T.O designation refers to an off-delay timer. True or false?
23. The electrical interlocks in a reversing magnetic starter are:
    A. Normally closed contacts
    B. Normally open contacts
    C. In parallel with the forward and reverse pushbuttons
    D. Controlled by the timing relays necessary for circuit operation

Answers

1. D
2. C
3. D
4. C
5. A
6. A
7. False
8. D
9. True
10. B
11. C
12. C
13. D
14. B
15. B
16. A
17. C
18. 3 and 4
19. True
20. Closed
21. True
22. False
23. A

Diagrams – Self Test 4

Questions

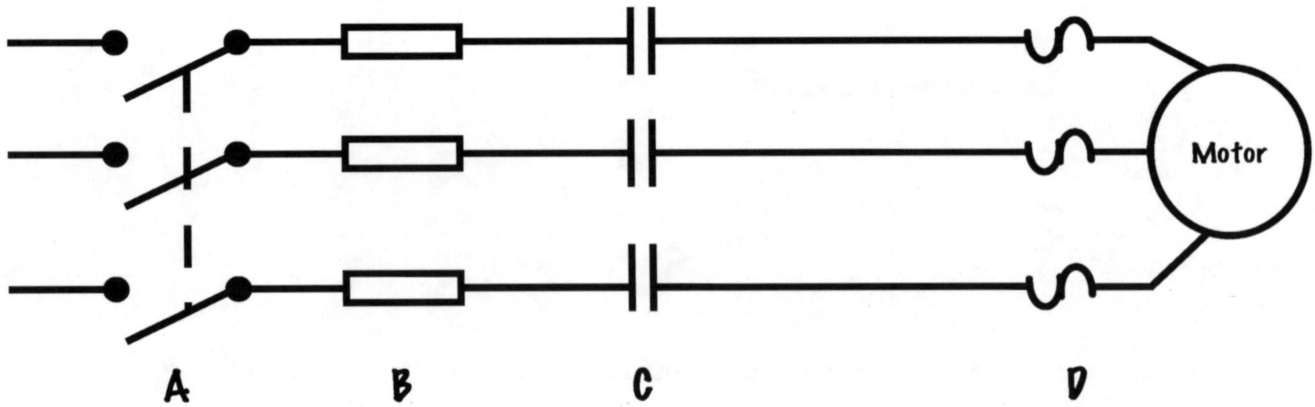

Using the above diagram, answer questions 1 to 5:

1. Which letter identifies the component that normally provides overcurrent protection for the motor branch circuit?
2. Which letter identifies the component that normally provides overload protection for the motor branch circuit?
3. When energized by the control circuit, which letter identifies the component that provides normal starting and stopping actions?
4. Between which two letters is the control circuit power normally taken from?
5. Which letter(s) identify devices that must be horsepower rated?

6. In a motor control circuit with multiple start/start pushbutton stations, the stop buttons would be connected in ______ and the start buttons would be connected in _______.
7. To troubleshoot an electrical control circuit, the best drawing to be used would be:
   A. Schematic
   B. Pictorial
   C. Wiring
   D. RIser
8. A drawing which shows the relative locations of various components is a:
   A. Wiring diagram
   B. Schematic diagram
   C. Elementary diagram
   D. Ladder diagram
9. Referencing the following drawing, what is the number of wires required where indicated?

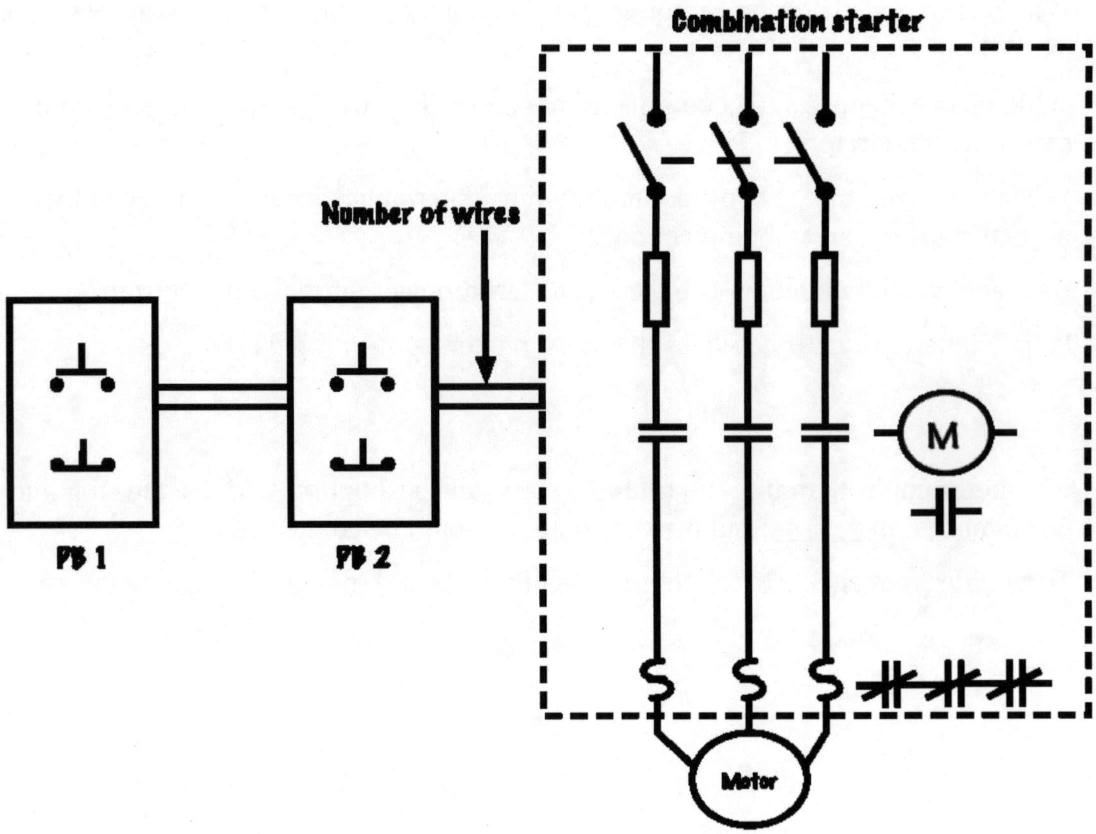

Answers

1. B
2. D
3. C
4. B and C
5. A and C
6. Series, parallel
7. A
8. A
9. 3

Circuits – Self Test 5

Questions

Using the following diagram, answer questions 1 to 5:

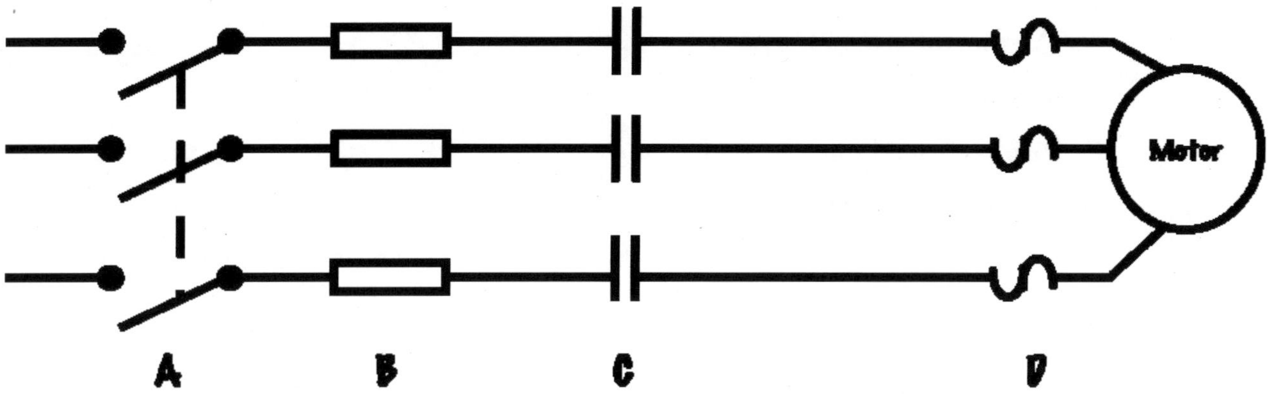

1. Which letter identifies the component that normally provides overcurrent protection for the motor branch circuit?
2. Which letter identifies the component that normally provides overload protection for the motor branch circuit?
3. When energized by the control circuit, which letter identifies the component that provides normal starting and stopping actions?
4. Between which two letters is the control circuit power normally taken from?
5. Which letter(s) identify devices that must be horsepower rated?

6. The minimum number of control circuit conductors to a stop/forward/reverse pushbutton station that provides low-voltage protection for a three-phase reversible motor is:
   - A. 2 wires
   - B. 3 wires
   - C. 4 wires
   - D. 6 wires
7. Inching is another term used for:
   - A. Jogging
   - B. Plugging
   - C. Shunting
   - D. Hunting
8. The rotation of an AC three-phase induction motor may be reversed by interchanging:
   - A. The control circuit
   - B. The forward/reverse interlocks
   - C. The forward reverse coils
   - D. Any two power lines

9. In a control schematic, a dotted line between two coils generally means that the two coils:
   A. Are operated together
   B. Are mechanically interlocked
   C. Are electrically interlocked
   D. Have a common set of contacts
10. The purpose of electrical interlocking in a three-phase reversing AC magnetic motor starter is to:
   A. Ensure that the forward direction of rotation is selected first
   B. Allow both coils to be energized together
   C. Prevent both coils from being energized simultaneously
   D. Maintain the coil circuit after the stop button is released
11. To connect a motor solely for jogging operation, the control circuit will:
   A. Use holding contacts
   B. Connect holding contacts in series with the start button
   C. Connect holding contacts in parallel with the start button
   D. Not use holding contacts
12. In regards to the drawing, the best label for the pushbutton labelled "Z" would be:
   A. Stop
   B. Jog
   C. Run
   D. Reset
13. In regards to the drawing, the best label for pushbutton "Y" would be:
   A. Stop
   B. Jog
   C. Run
   D. Reset

Use the following image to answer questions 14 and 15.

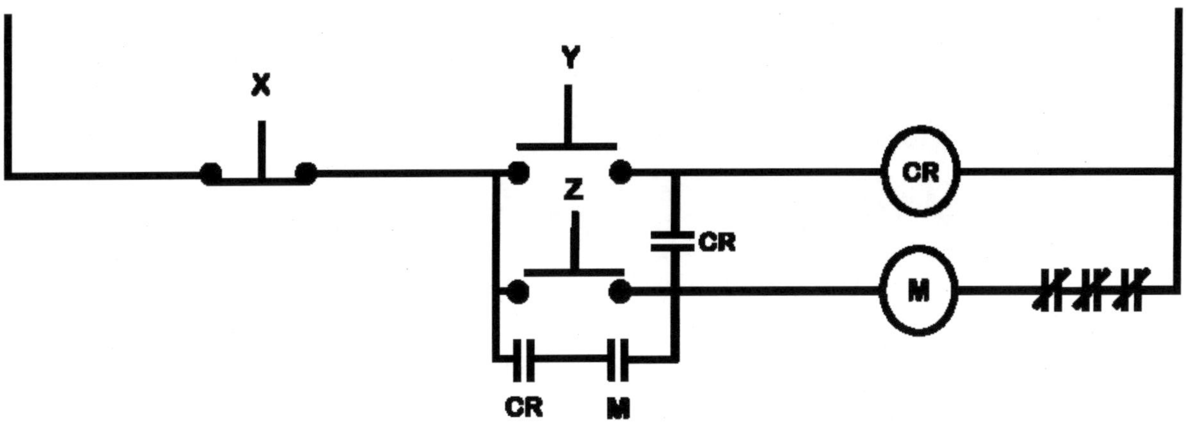

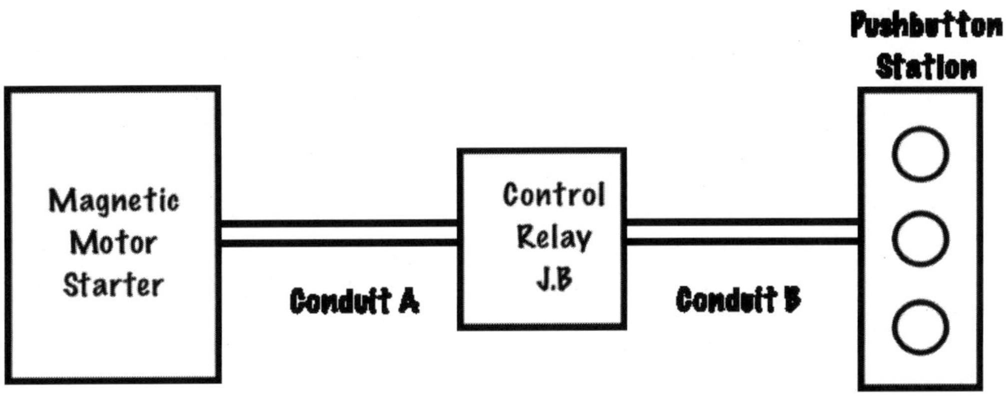

14. What is the minimum number of control circuit conductors required in Conduit A?
    - A. 2 wires
    - B. 3 wires
    - C. 4 wires
    - D. 5 wires
15. What is the minimum number of control circuit conductors required in Conduit B?
    - A. 2 wires
    - B. 3 wires
    - C. 4 wires
    - D. 5 wires

16. Timed relays are the best method to provide plugging for a motor. True or false?
17. A time-delay relay includes both timed and instantaneous contacts. True or false?

18. A N.C.T.C contact will be open immediately after the coil is de-energized. True or false?
19. A N.O.T.O contact will be closed while the coil of the relay is energized. True or false?
20. A N.C.T.O designation refers to an off-delay timer. True or false?
21. The electrical interlocks in a reversing magnetic starter are:
    A. Normally closed contacts
    B. Normally open contacts
    C. In parallel with the forward and reverse pushbuttons
    D. Controlled by the timing relays necessary for circuit operation
22. Overload protection for motors is achieved by connecting:
    A. Thermal sensing elements in series with the motor
    B. A reset relay in series with the motor
    C. Cartridge fuses in series with the motor
    D. A set of N.C contacts in series with the motor
23. Jogging refers to a:
    A. Motor that is not capable of developing constant torque
    B. Motor that is polyphasing
    C. Motor that periodically starts and stops
    D. Method used to stop a motor for exact positioning

Answers

1. B
2. D
3. C
4. B and C
5. A and C
6. C
7. A
8. D
9. B
10. C
11. D
12. B
13. C
14. ?
15. C
16. False
17. True
18. False
19. True
20. False
21. A
22. A
23. D

Troubleshooting – Self Test 6

1. A drawing which shows the relative locations of various components is a:
    - A. Wiring diagram
    - B. Schematic diagram
    - C. Elementary diagram
    - D. Ladder diagram
2. To ensure safety for maintenance, the motor disconnecting switch must be locked in the "OFF" position. After the maintenance work is completed, the lock is removed by:
    1. The supervisor
    2. The project manager
    3. The person who put the lock on
    4. The lead hand
3. With regard to workers safety, "isolation" refers to:
    1. Moving to a remote location
    2. Disconnect from all energy sources
    3. Turn off the electrical switch
    4. Fence off the work site
4. Referring to the drawing, assuming that fuse C is blown, which pair of points will result in a reading of zero volts when the voltmeter is connected across them?

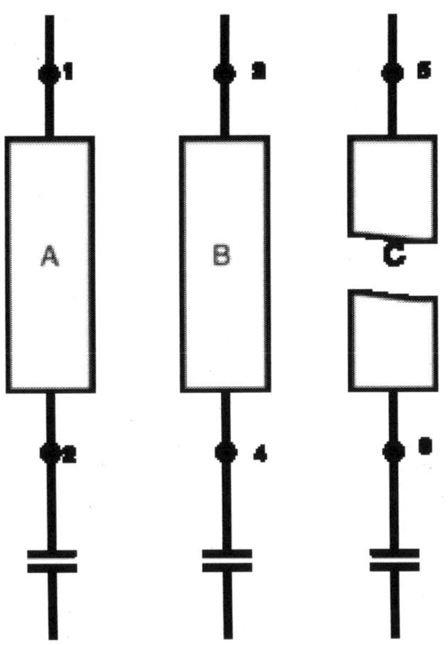

5. The disconnecting means which is NOT intended for interrupting current flow is the:
    1. Motor switch
    2. General use switch

3. Isolating switch
4. Circuit breaker

6. If the circuit in the drawing was operating normally and an overload occurred, then:

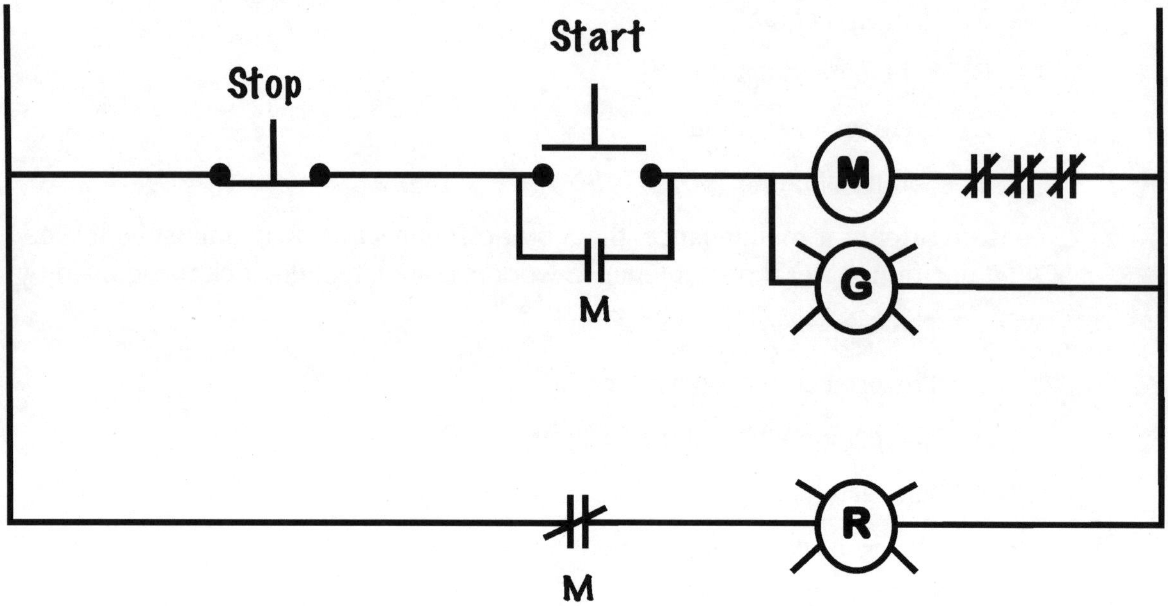

A. Both lights would come on
B. The green light would come on, and the red light would go out
C. Both lights would go out
D. The red light would come on, and the green light would go out

7. During normal operation, a loud chattering noise is heard from the enclosure of an AC magnetic starter. What is the most likely cause?

A. Broken shading coil
B. Open in the seal-in circuit
C. Rust on pole faces
D. Power contact is not making good contact due to poor pressure

8. If the control circuit shown below tripped due to overload, then which of the voltmeter positions shown would indicate line voltage?

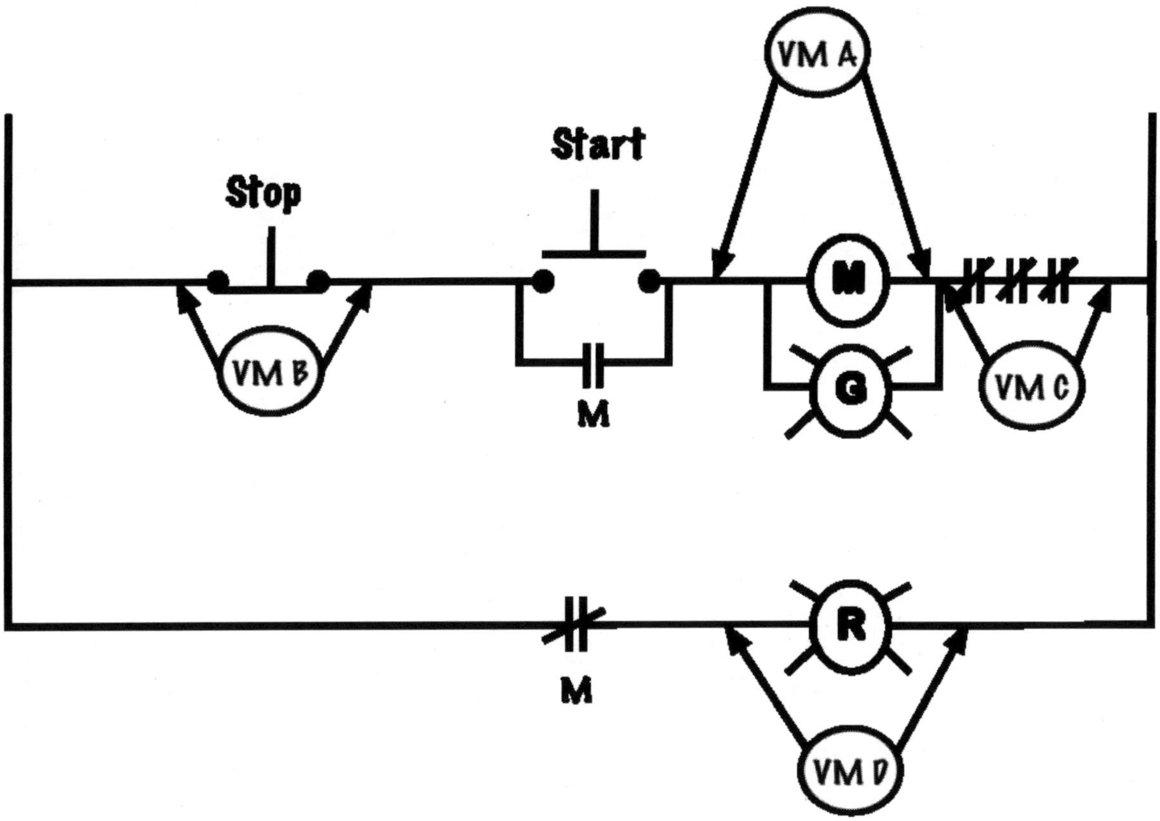

1. VM A
2. VM B
3. VM C
4. VM D

9. What is the minimum number of control circuit conductors required in Conduit A?

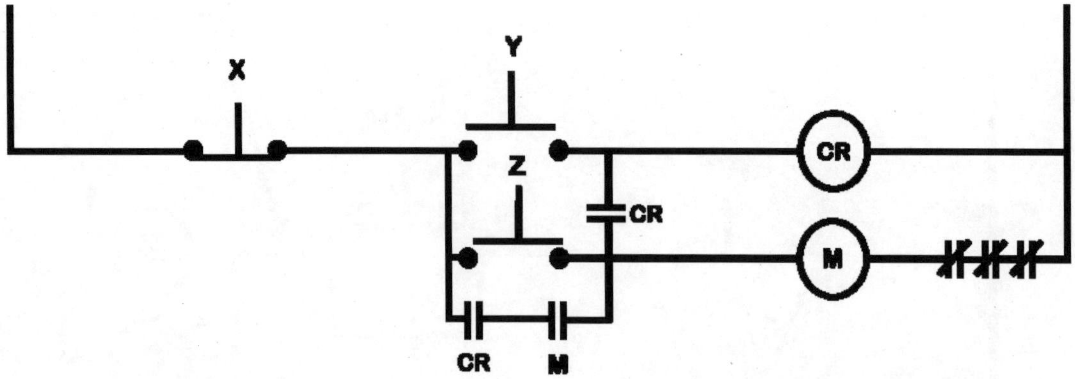

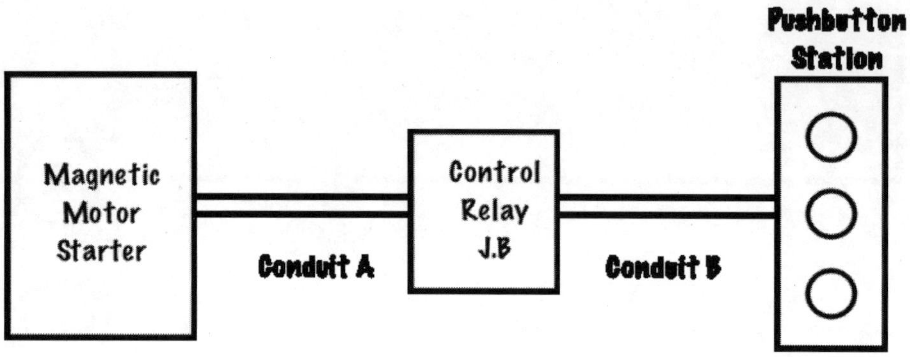

A. 2 wires
B. 3 wires
C. 4 wires
D. 5 wires

10. An isolating switch can be used as a motor power circuit disconnect switch. True or false?

Use the following image to answer questions 11 and 12.

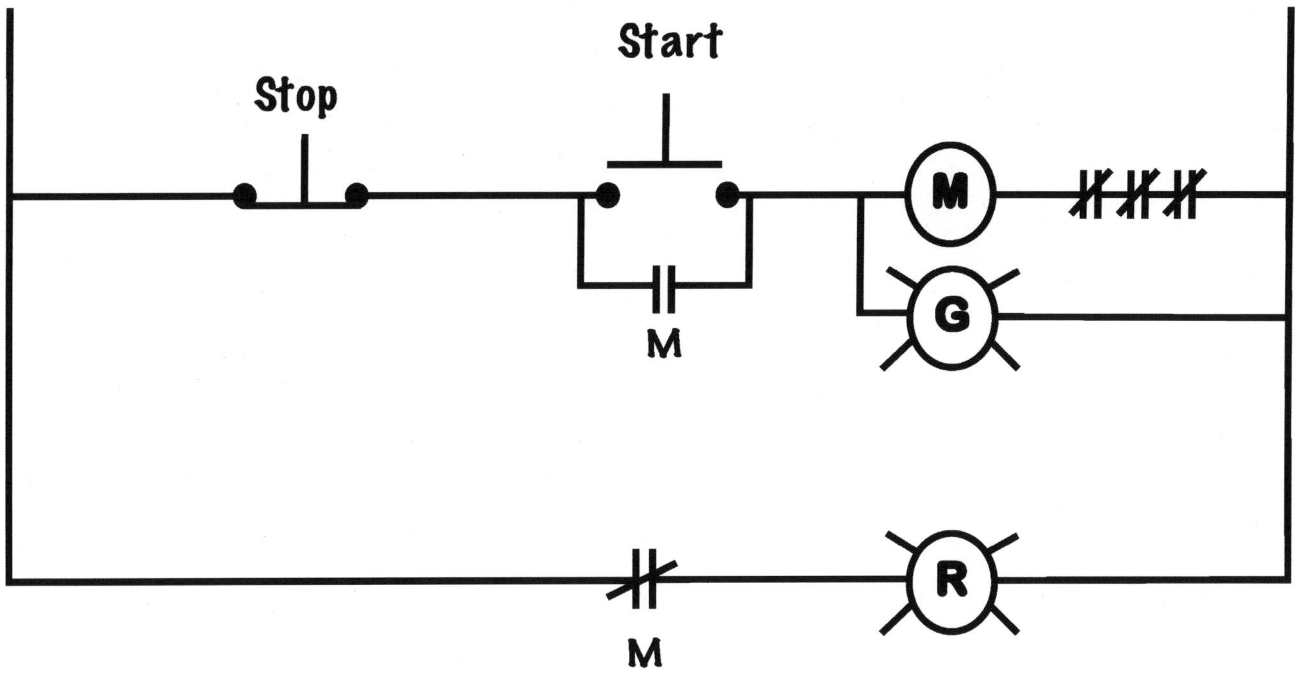

11. If the motor starter in the drawing is energized and functioning properly, what should the voltage be across the normally open contact (M)?
    - A. Line voltage
    - B. Zero voltage
    - C. Half of line voltage
    - D. Twice line voltage
12. If the motor starter in the drawing is energized and functioning properly, what should be the voltage across the N.C contact (M)?
    - A. Line voltage
    - B. Zero voltage
    - C. Half of line voltage
    - D. Twice line voltage

Answers

1. A
2. C
3. B
4. 3 and 6
5. C
6. D

7. A
8. ?
9. ?
10. False
11. B
12. A

Glossary

2-3 Holding Contact

Also known as a "maintaining" contact, these are the **normally open** contacts of a magnetic starter that are connected in parallel with the start button in a three-wire control circuit. When using the conventional NEMA numbering system, they get wire numbers "2" and "3."

Ammeter

A device used for measuring the value of current flowing through a circuit. The ammeter has a very small internal resistance and so must be connected in series with the circuit to prevent dangerous levels of current from flowing.

Can be digital or analogue and measure either AC or DC.

Ampere (A)

The unit used to measure electrical current. It is equal to a flow of one coulomb per second. It may also be called "amp."

Anti-Plugging Circuit

A circuit that is not designed to suddenly stop or reverse a motor until the shaft has come to a rest.

Armature

With respect to magnetic contactors, the armature or plunger is the movable part of the magnetic circuit. When a coil is energized the armature is pulled in, opening and/or closing a set or sets of contacts.

Auxiliary Contacts

Contacts on a magnetic starter that are not Horsepower rated. Can come as either normally-open or normally-closed and can be used as maintaining contacts, electrical interlocks or control for pilot lights.

Circuit Breaker

An automatic device that is designed to safely disconnect circuits under fault conditions. Most circuit breakers provide Overload and Overcurrent protection, and are rated in Volts, Amps and Horsepower.

Circuit Diagram

A diagram that uses lines and symbols to show the electrical continuity and connections of a circuit. Can be used either for troubleshooting (schematic diagrams) or connection purposes (wiring diagrams).

Contact

The conducting part of a switch that makes or breaks a circuit.

Continuous-Duty Rating

The maximum rated value of RMS current that the overcurrent device is designed to handle on a continuous basis without tripping.

Control Circuit

In contrast to the Power Circuit, the Control Circuit consists of inputs, in the form of switches, pushbuttons or pilot devices, which when activated, can either directly, or through a magnetic motor starter, energize a load. The Control Circuit often operates at a lower voltage than the Power Circuit for safety and ease of installation.

Control Relay

Used when additional auxiliary contacts are needed in a control circuit, a control relay is a magnetic contactor which is **not** designed for the energization of motors, and **does not** have built in overload protection.

Crisscross Voltmeter Method

The crisscross voltmeter method is used to troubleshoot fuses on a live circuit. It involves checking the potential difference between the line and load sides of fuses connected to different phases in a three-phase fusible disconnect.

Current (I)

The rate of flow of an electric charge, measured in amperes (or amps). When one coulomb of charge moves past one point in once second, current is said to flow at a rate of one ampere. Current flows from negative potential to a positive potential through a load.

Electrical Interlock

Normally-closed contacts used in forward/reverse control circuits that prevent both directions coils from being energized at the same time.

Electrically Common

Referring to two or more points in a circuit which have no loads or switches between them and have no potential difference between them.

Fuse

An insulated tube containing a strip of conductive metal that has a lower melting point than either copper or aluminum. It protects a circuit from damage because it will melt in overload or overcurrent situations and break the connection with the rest of the circuit.

Horsepower-Rated Contact

A set of contacts that has been designed to make or break the flow of current to a motor and can handle loads up to a specified amount of horsepower at a specific voltage. The size of contacts are determined by the size of the motor they are controlling.

Inching

The term given to the momentary energization, at *reduced* voltage, of a motor only so long as an operator is pressing a button.

Inrush Current

The initial high value of current produced when an inductive load is first energized.

Interrupting-Capacity (IC) Rating

The maximum fault current that an overcurrent device can interrupt without damage to itself. Most circuit breakers and fuses have an IC rating of 10,000 amps.

Isolating Switch

A switch that is used to lockout an electrical circuit once a motor has been turned off to ensure it is completely de-energized. Not designed to interrupt the flow of current.

Jog Circuit

A control circuit that allows an operator to either start the motor or "Jog" the motor only so long as a pushbutton is depressed, and are commonly used for motors controlling conveyor belts to allow for precise positioning of materials.

Jogging

The term given to the momentary energization, at *full* voltage, of a motor only so long as an operator is pressing a button.

Kirchhoff's Current Law

"The sum of the currents entering a point must equal the sum of the currents exiting that point."

In a series circuit the same value of current flows through each device and the source.

I total = I1 = I2 = I3...

In a parallel circuit, each individual branch will contribute its current to the total current.

I total = I1+ I2 + I3...

Kirchhoff's Voltage Law

"The sum of the voltage rises must equal the sum of the voltage drops in a circuit."

In a series circuit, the total voltage at the source is equal to each of the individual voltage drops in any loads.

E total = E1 + E2 +E3...

Loads connected in a parallel circuit will experience the same potential difference.

E total = E1 = E2 = E3...

Ladder Diagram

See Schematic Diagram.

Locked-Rotor Current

The current drawn by a motor when the motor is not spinning.

Lockout Relay

A special solenoid coil included in a plugging switch assembly that disables the contacts until the motor itself has been energized.

Low-Voltage Protection (LVP)

Circuits with low-voltage protection will not automatically turn back on when voltage is restored following a power outage. Examples include the microwave or power tools.

Low-Voltage Release (LVR)

Circuits with low-voltage release are designed to re-energize automatically when voltage is restored after a power outage. Examples include lights or the kitchen fridge.

Mechanical Interlock

A physical barrier that is pushed into the path of one coil's armature by the movement of the adjacent coil in a forward/reversing motor starter.

Motor Starter

A device that controls the flow of electrical power to a motor. It is designed to safely start and stop a motor, and provide **overload protection**.

Normally-Closed (NC) Contact

A contact that under normal conditions has continuity through it. When the contact changes its state it interrupts the flow of current by opening its contacts. Can be associated with pushbuttons, pilot devices or magnetic contactors.

Normally-Open (NO) Contact

A contact that under normal conditions does not have continuity through it. When the contact changes its state it permits the flow of current by closing its contacts. Can be associated with pushbuttons, pilot devices or magnetic contactors.

Numbering System

Used to convert between wiring and schematic diagrams, the numbering system is a method of identifying and labelling each electrically common point in a circuit. Two wires are considered electrically common if they share an electrical connection with no switches or loads between them, and so would be assigned the same number in the diagram.

Ohm

The unit used to measure electrical resistance (Ω). It takes one volt to push one amp through one ohm of resistance.

Ohm's law

Current = Voltage divided by Resistance (or I=E/R).

Ohmmeter

A device used to measure the resistance of a circuit. Ohmmeters must not be used on live circuits. Ohmmeters connect a small internal voltage source to the circuit that is being measured or tested, and determine the value of resistance or continuity by measuring what value of current flows through the meter.

Can be either digital or analogue.

Overcurrent

A sharp and fast rise in current over a short period of time (fractions of a second) where the value of current is far greater than the nominal line current.

Overload

A moderate and gradual rise in the value of current over a relatively long period of time that is caused by excessive amounts of current drawn by a motor due to too much load being put on the motor.

Overload Relay (OLR)

A heater element paired with normally-closed contacts that open once the heater gets too hot. Two types of relays are the bimetallic strip and the melting solder pot.

Parallel

In electrical terms, refers to a connection where current has more than one path to flow.

Loads connected in parallel will experience the same potential difference (voltage), but may draw different values of current depending upon their individual resistance.

Phase Rotation

The direction that a three-phase motor spins is determined by the phase sequence of the voltage impressed upon it. To reverse the direction of the motor we simple reverse the phase sequence by switching any to line leads.

Pilot Device

An auxilary device that provides indication or control of a process to an operator. Pilot devices include automatic switches such as float and pressure switches, as well as indicating lights.

Pilot Light

A small lamp connected in the control circuit to indicate the status of a motor or other situation.

Plugging

When a motor is spinning in one direction and is stopped and suddenly re-energized in the opposite direction before the shaft of the motor has time to come to a complete stop.

Plugging Circuit

A circuit that utilizes the momentary reversal of the direction of a three-phase motor to bring it to a sudden stop. Can be very hard on the motor and any driven equipment.

Power

The rate at which work is done. It is measured in **watts (W)**, or **joules per second (J/s)**.

Power Circuit

In contrast to the control circuit, the power circuit provides the large values of voltage and current used by the motor itself. Must be equipped with overcurrent and overload protection, and horsepower-rated contacts in the control gear equal to the voltage and current ratings of the motor.

Process Variable

A measured value of a particular part of a process that is being monitored or controlled, such as temperature, pressure, liquid level, flow, position or proximity.

Pushbutton

A momentary contact device that has a built in spring to return the button to its normal position once release. Available with either normally-open, normally-closed or both sets of contacts.

Resistance (R)

The opposition to the flow of current in an electric circuit, measured in ohms (Ω).

Schematic Diagram

A diagram that shows how a circuit works logically and electrically. It uses symbols to identify components and interconnecting lines to display the electrical continuity of a circuit. It is often used for troubleshooting purposes. Also known as a ladder diagram.

Series

In electrical terms, refers to a connection where current has only one path to flow.

Loads connected in series will have the the same value of current flowing through them, and share the total voltage between them. Switches and overcurrent equipment is connected in series with equipment to control and protect it.

Single-Phase System

The simplest electrical circuit. It requires only two lines: one for power to go in and the other is a return path for current. These are often called Line 1 and Line 2, or Line 1 and Neutral. Current only has one path to travel in a single-phase circuit, such as a control circuit.

Switch

A device for making or breaking the connection in an electric circuit.

Three-Phase System

An electrical circuit that uses three current carrying conductors, called Line 1, Line 2, and Line 3, which have a 120° phase shift in the voltage and current waveforms between them. The Power Circuit of three-phase motors is an example of a three-phase circuit.

Three-Wire Circuit

In motor control terminology, a three-wire circuit utilizes a **magnetic motor starter** with a holding contact, along with momentary contact pushbuttons. A three-wire circuit provides **low-voltage-protection**.

Time-Delay Relay

A **control relay** equipped with a delay mechanism that can be used to alter the timing effects of a set of contacts. There are two varieties of timer relays: **On-delays** which have their delay action when they are energized and revert to their original condition instantly upon de-energization, and **off-delays**, which change the state of their contacts instantly upon energization, but have a delay before reverting to their normal condition upon de-energization.

Two-Wire Circuit

In motor control terminology, a two-wire circuit utilizes a **manual motor starter** with a maintained contact. A two-wire circuit provides **low-voltage release**.

Voltage (E)

The difference in electric potential between two points, which is defined as the work needed per unit of charge to move a test charge between the two points. It is measured in volts (V).

Voltage Rating

The maximum amount of voltage that a fuse, circuit breaker, switch-gear or motor starter can handle. The voltage rating of a fuse or circuit breaker must be equal to or greater than the system voltage.

Voltmeter

A device testing and measuring the potential difference (voltage) between two points. Leads are connected in parallel with the circuit, and the meters very high internal resistance will draw a small current which can be used to determine the level of voltage.

Can be digital or analogue and measure either AC or DC.

Watt (W)

The unit used to measure power in an electric circuit, equivalent to **one joule per second**, or the power dissipated when one **volt** pushes one **amp** through a circuit.

Watt's law

Power = Voltage times Current (or W=EI)

Wiring Diagram

A diagram shows how equipment is laid out and the connections between them. This type of diagram shows the physical relation of all devices in the system, the conductor terminations between these devices, and are commonly used in motor control installations. Also known as a connection diagram.

Zero-Speed Switch

A switch that is activated by centrifugal forces and is attached either directly or via a belt to the shaft of the motor.

Index